book is to be returned on or before

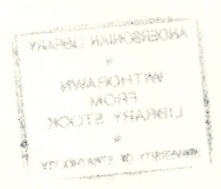

An international initiative by the
World Federation for Culture Collections,
with financial support from UNESCO

LIVING RESOURCES FOR BIOTECHNOLOGY

Editorial Board:

A. Doyle
D. L. Hawksworth
L. R. Hill
B. E. Kirsop (Senior Editor)
K. Komagata
R. E. Stevenson

Yeasts

LIVING RESOURCES FOR BIOTECHNOLOGY

Yeasts

Edited by

B. E. Kirsop and C. P. Kurtzman

in collaboration with
T. Nakase and D. Yarrow

The right of the
University of Cambridge
to print and sell
all manner of books
was granted by
Henry VIII in 1534.
The University has printed
and published continuously
since 1584.

CAMBRIDGE UNIVERSITY PRESS

Cambridge
New York   New Rochelle   Melbourne   Sydney

Published by the Press Syndicate of the University of Cambridge
The Pitt Building, Trumpington Street, Cambridge CB2 1RP
32 East 57th Street, New York, NY 10022, USA
10 Stamford Road, Oakleigh, Melbourne 3166, Australia

First published 1988

Printed in Great Britain at the University Press, Cambridge

British Library cataloguing in publication data
Yeasts. – (Living resources for
biotechnology).
1. Yeast fungi
I. Kirsop, B. E.   II. Kurtzman, C. P.
III. Series
589.2'33   QK617.5

Library of Congress cataloguing in publication data
Yeasts.
(Living resources for biotechnology)
Bibliography: p.
Includes index.
1. Yeast fungi – Biotechnology.   I. Kirsop, B. E.
II. Kurtzman, C. P.   III. Nakase, T.   IV. Yarrow, D.
V. Series.
TP248.27.Y43Y42   1987   660'.62   87–27646

ISBN 0 521 35227 4

wv

P
660.62
LIV

CONTENTS

CONTRIBUTORS

Allner, K. Public Health Laboratory Service Centre for Applied Microbiology and Research, Porton Down, Salisbury, Wiltshire SP4 0JG, UK (Chapter 3)

Allsopp, D. CAB International Mycological Institute, Ferry Lane, Kew, Surrey TW9 3AF, UK (Chapter 7)

Bousfield, I. J. National Collections of Industrial and Marine Bacteria Ltd, Torry Research Station, 135 Abbey Road, PO Box 31, Aberdeen AB9 8DG, UK (Chapter 6)

DaSilva, E. J. Division of Scientific Research and Higher Education, United Nations Educational Scientific and Cultural Organisation, 7 Place de Fontenoy, 75700 Paris, France (Chapter 8)

Fabricius, B.-O. Department of Microbiology, University of Helsinki, SF-00710 Helsinki, Finland (Chapter 2)

Kirsop, B. E. Microbial Strain Data Network, Institute of Biotechnology, Cambridge University, 307 Huntingdon Rd, Cambridge, CB3 OJX, UK (Chapters 1, 3, 4, 8)

Krichevsky, M. I. Microbial Systematics Section, Epidemiology and Oral Disease Prevention Program, National Institute of Dental Research, Bethesda, Maryland 20892, USA (Chapter 2)

Kurtzman, C. P. Northern Regional Research Center, Agricultural Research Service, US Department of Agriculture, Peoria, Illinois 61604, USA (Chapters 1, 3, 5)

Nakase, T. Japan Collection of Microorganisms, The Institute of Physical and Chemical Research, Hirosawa, Wako-shi, Saitama 351, Japan (Chapters 1, 3, 4, 5)

Simione, F. P. American Type Culture Collection, 12301 Parklawn Drive, Rockville, Maryland 20852, USA (Chapter 7)

Sugawara, H. Life Science Research Information Section, RIKEN, Wako, Saitama 351-01, Japan (Chapter 2)

Yarrow, D. Centraalbureau voor Schimmelcultures, Yeast Division, Julianalaan 67A, 2628 BC Delft, The Netherlands (Chapters 1, 3, 4, 5)

SERIES INTRODUCTION

The rapid advances taking place in biotechnology have introduced large numbers of scientists and engineers to the need for handling microorganisms, often for the first time. Questions are frequently raised concerning sources of cultures, location of strains with particular properties, requirements for handling the cultures, preservation and identification methods, regulations for shipping, or for the deposit of strains for patent purposes. For those in industry, research institutes or universities with little experience in these areas, resolving such difficulties may seem overwhelming. The purpose of the World Federation for Culture Collections' (WFCC) series, Living Resources for Biotechnology, is to provide answers to these questions.

Living Resources for Biotechnology is a series of practical books that provide primary data and guides to sources for further information on matters relating to the location and use of different kinds of biological material of interest to biotechnologists. A deliberate decision was taken to produce separate volumes for each group of microorganism rather than a combined compendium, since our enquiries suggested that inexpensive specialised books would be of more general value than a larger volume containing information irrelevant to workers with interests in one particular type of organism. As a result each volume contains specialised information together with material on general matters (information centres, patents, consumer services, the international coordination of culture collection activities) that is common to each.

The WFCC is an international organisation concerned with the establishment of microbial resource centres and the promotion of their activities. In addition to its primary role of coordinating the work of culture collections throughout the world, the committees of the WFCC are active in a number of areas of particular relevance to biotechnology, such as patents, microbial information centres, postal and quarantine regulations, educational and conservation matters (see Chapter 8). The Education Committee of the WFCC proposed the preparation of the current volumes.

The WFCC is concerned that this series of books is of value to biotechnologists internationally, and the authors have been drawn from specialists throughout the world. The close collaboration that exists between culture collections in every continent has made the compilation of material for the books a simple and pleasurable process, since the authors and contributors are for the most part colleagues. The Federation hopes that the result of their labours has produced valuable source books that will not only accelerate the progress of biotechnology, but will also increase communication between culture collections and their users to the benefit of both.

Barbara Kirsop
President, World Federation
for Culture Collections

PREFACE

Yeasts have been used for centuries in the making of bread and fermented drinks, and it is only in the last fifty years that their potential for other applications has been realised. More recently the developments in genetics and molecular biology have opened up the possibility of the genetic manipulation of yeasts for commercial purposes.

These simple fungi are a phylogenetically diverse group of organisms, and early taxonomic studies established their physiological differences which, together with morphological characteristics, were used to differentiate them. Although this heterogeneity increases the industrial capabilities of yeasts, it also follows that a number of methods are required for their growth and general manipulation. Additionally, protocols necessary for their long-term preservation can vary, as genera, species and even strains show differing resistances to the stresses imposed by preservation procedures.

Their identification depends on both morphological and physiological examination, and expert help is generally required to interpret test results. Although the standard taxonomic works on yeasts, such as editions 2 and 3 of *The Yeasts: A Taxonomic Study* edited by J. Lodder and N. J. W. Kreger-van Rij, respectively, give basic guidance on yeast culture, preservation, and identification, their main purpose is to describe accepted yeast species. These descriptions provide a wealth of information on the physiology and sexual life cycles of species that may have direct industrial application.

The present volume seeks to bring into one book information, or sources of information, relating to the use of yeast in a biotechnological context. Thus, the first chapter introduces the major service collections of yeasts and describes the holdings and services they provide. These collections are not only resource centres for cultures, but also, increasingly, information centres; these, as well as regional and international centres concerned only with data rather than the supply of microorganisms, are outlined in Chapter 2. The administrative practices adopted by culture collections and essential information on the health and safety aspects of microbiology are described in Chapter 3, while Chapters 4 and 5 cover in some detail procedures used for the growth, preservation and identification of yeasts.

More commercial aspects of biotechnology, such as the requirements for patenting and the sources and kinds of microbiological contract work available to biotechnologists, are described in Chapters 6 and 7. The final chapter provides information on the organisation of microbial resource centres throughout the world, and it highlights the work of their federations and committees. These are concerned with the co-ordination of culture collection activities regionally and internationally, the development of new initiatives, training courses and general educational matters and, perhaps of greatest importance, the conservation of the microbial gene pool and the protection of endangered species.

This brief introduction touches on only a few of the topics covered in this volume. We believe that the book addresses the main questions asked by those concerned with the use of yeast cultures and acts as a pointer to other sources of information. We hope it will encourage readers to establish closer links with culture collections, the source of so much information on subjects directly related to biotechnology. Above all, we hope those making their first contact with the study of yeasts will find it to be as fascinating and rewarding as do the authors.

B. E. Kirsop
C. P. Kurtzman
May 1987

ACKNOWLEDGEMENTS

We would like to thank our colleagues from many parts of the world who have contributed to the material found in this book. In particular, we are very grateful to the curators of the yeast collections for so willingly responding to our requests for information. Without the friendly co-operation and generosity that exists between culture collection workers throughout the world, the difficulties in preparing this book would have been insurmountable.

1

Resource centres

B. E. KIRSOP

1.1 Yeasts as tools for biotechnology

Microfungi multiplying predominantly by budding are collectively called yeasts. The term has no systematic significance, since budding microfungi are found in diverse groups of fungi, and 'yeasts' are frequently understood by the general public to refer to the species responsible for brewing and baking, *Saccharomyces cerevisiae*. There are, however, over 460 species of yeasts, and their different characteristics and biotechnological applications are discussed in Chapter 5.

Because of their long-established association with industry, yeasts have been well-studied physiologically and genetically and it has become recognised that they represent a group of reactive, robust, genetically manipulable microorganisms. Their tolerance to a wide range of temperatures, levels of pH, oxygen, salt and sugar; their sexual life cycle; their ability to grow on a wide range of substrates; and their ease of handling in bulk have stimulated interest in their potential role in biotechnology. More recently the discovery and characterisation of plasmids in a number of species has focused attention on the possibility of constructing strains with unique characteristics.

Yeasts have been used for brewing, baking and wine making for many centuries and must represent the earliest biotechnological agents. As a result of their long history in the service of man much has been learned about ways of controlling their activity, and this knowledge has enabled their use to be developed from the domestic scale to a cottage industry and hence, in more recent times, to full commercial scale. Over the last few decades attention has been directed particularly to their use for the production of enzymes, medically important substances such as insulin and interferon, protein for animal or human food and alcohol for fuel.

1

Research continues into the study of the physiological and genetic properties of yeasts and in particular their use in the construction of vectors for the modification of microorganisms able to produce novel and valuable gene products.

1.2   Resource centres for yeasts

There exists throughout the world a large number of culture collections that include yeasts in their holdings. Most, but not all, are listed in the second edition of the World Federation for Culture Collections' *World Directory of Collections of Cultures of Microorganisms* (V. F. McGowan and V. B. D. Skerman, 1982) available from the Secretary, UNEP/UNESCO/ICRO Panel on Microbiology, Swedish University of Agricultural Sciences, S-750 07 Uppsala, Sweden in book form or microfiche. (The third edition (1986) may be obtained from the United Nations Environment Programme, Information Service, PO Box 30552, Nairobi, Kenya.) The collections in the Directory vary substantially in their composition and activities, depending on the reason for their establishment, and the different categories of collections are described below.

1.2.1   *Service culture collections*

A number of collections exist primarily to provide cultures and services to the scientific community. They are supported in the main from public funds and are often described as national or regional collections. Many are long-established; others have been formed recently or developed from earlier specialised collections often in direct response to advances in biotechnology. In general, cultures are provided on demand and without discriminating between category (education, research, industry) or nationality of user. Almost all make a charge for the cultures and services they provide, although different levels of charges may be levied and a distinction made between industrial and non-industrial users.

Functions of service culture collections. The collection, preservation and supply of cultures of microorganisms are essential functions of all service culture collections. They cannot fulfil the role of resource centres for the future unless they are actively engaged in the collection and conservation of microbial genetic material, and collection staff must scan current literature and contact scientists using cultures of potential interest to others in order to increase their holdings. Stocks of cultures are also built-up through the work of the collections themselves (identification, screening programmes, contract projects), by exchange with other collections and by the voluntary deposit of strains by other scientists.

The type of material conserved may be broad-ranging – a taxonomically comprehensive collection, for example – or specialised, perhaps only holding organisms from a marine environment or genetic strains.

The preservation of strains by techniques that lead to maximum survival levels and strain stability occupies much of the attention of service collections, since it is essential that cultures continue to represent the original deposits after prolonged storage. Methods commonly used by yeast culture collections are described in Chapter 4, and research continues into improving preservation procedures, particularly for sensitive strains.

The supply of cultures is a function of service collections that distinguishes them from other types of collections. The income from the sale of cultures is used partially to off-set the costs of maintaining the microbial gene pool, and the charges made and principles adopted for arriving at these are discussed in Chapter 3. Because of the reusable nature of the product it is not possible to cost cultures in the same way as, say, chemicals, since above an 'acceptable' level of charges a 'black market' begins to operate and microbiological standards fall as unauthenticated or imperfectly preserved material is used.

In addition to the three essential functions (collection, preservation, supply) common to all service collections, other services may be provided. Most major collections offer identification services, and may in addition be able to provide preservation, safe deposit, patent depository, contract work (see Chapter 7) and training services. Services available from yeast centres are listed in Table 1.4. Charges are again made for these services to further off-set costs.

Without exception, all the service collections listed in section 6 have their own programme of research. This not only ensures close contact with current research activities in other laboratories but advances the acquisition of information essential to the proper functioning of the collections. Research is generally directed towards taxonomy, identification or preservation technology. Additionally, work may be carried out to obtain further information on the properties of maintained strains. This may relate to properties such as environmental tolerances, genetical characterisation of plasmids, industrial performance or drug resistance, all data of particular interest to biotechnologists.

1.2.2   *In-house collections*

Many collections are set-up for the purpose of providing a service to the organisation in which they are housed. The ARS Culture Collection, Northern Regional Research Center (see Section 1.6.4), for example, exists primarily as a research and service laboratory to the US Department of Agriculture and functions as a service to the microbiological community only secondarily. Other in-house collections exist within industry, providing cultures and general microbiological support to company laboratories. Some of these collections are among the biggest, often containing large numbers of strains isolated during screening programmes. Occasionally companies are prepared to make some of their cultures or information about them available to others if this is not to their commercial disadvantage.

In-house collections do not usually prepare a catalogue for public use and do not necessarily provide cultures and services on demand to microbiologists that are not in their organisation. They nevertheless form an important and often unique microbiological resource.

1.2.3   *Research collections*

Private collections often develop as a result of the activities of an individual research worker with a specialised interest. Microorganisms maintained as a result of research activities are frequently very well characterised, although often in a narrow area, and form an essential scientific adjunct to publications. As such they should be conserved and made accessible to other microbiologists in the future. As long as the original worker continues to carry out research in this area the cultures are not endangered, but when a change of interest, change of job, or retirement occurs, the future preservation of the cultures may become of low priority and they should be transferred to one of the service collections for future safekeeping. Ideally, transfer of valuable material should occur while the research workers are still active in the area so that the maximum transfer of associated data can take place. Some scientific journals are taking positive steps to encourage the deposit of strains quoted in publications into a recognised culture collection. The more widespread adoption of such a policy would ensure that important strains become part of the microbiological heritage, and that published work is capable of further study by others.

1.2.4   *Data centres*

Data about microorganisms as well as the cultures themselves form essential resources for biotechnologists, and a number of centres have been set up to establish improved access to the data. These centres may be associated with culture collections or sited independently. They are an important and growing asset to microbiologists and are described in detail in Chapter 2.

1.3   Accessing the centres

Faced with the need for cultures or data about them, biotechnologists may be uncertain whom to contact for help. The following organisations exist to provide the information needed, and all but the last two are described further in Chapters 2 and 8.

> World Federation for Culture Collections (WFCC)
> WFCC's World Data Center on Collections of Microorganisms (WDC)
> Microbial Strain Data Network (MSDN)
> Microbial Resource Centres (MIRCENs)
> Regional organisations (e.g. European Culture Collections' Organization (ECCO) in Europe)
> National federations for culture collections
> Biotechnology data centres
> Individual culture collections/catalogues
> Microbiological societies

Lists of individual service collections follow in Section 1.6. Lists of microbiological societies may be obtained from the directory of the International Union of Microbiological Societies (IUMS). The current address of the General Secretary of IUMS may be obtained from the Secretariat of the International Council of Scientific Unions (ICSU), 51 Boulevard de Montmorency, 75016 Paris, France; Telephone (331) 525 03 29; Telex ICSU 630553 F.

There are a large number of commercial and publicly funded biotechnology databases often providing information about culture collections. Information about them may be obtained by application to national associations for biotechnology, the names and addresses of which are given in Chapter 8.

1.4   Future development of resource centres

Biotechnology has had a substantial impact on the development of culture collections. Throughout the world international and national funding agencies and government departments are assessing existing resources, building new centres, expanding and rationalising present ones and generally considering the level of commitment required for the future.

The demands made by the microbiological community on culture collections are growing and now include not only the collection, maintenance and supply of cultures, but also the development of 'centres of expertise', for studies in taxonomy, identification, preservation, computer applications and basic skills necessary for the culture of specialised groups of microorganisms (animal cells, methanogens, extreme thermophiles, for example). Additionally, there is a need for centres to provide associated services for the support of industry (biodegradation, isolation, precompetitive screening), and these are described further in Chapter 7.

It has been recognised over the past few years that there is a need for depositories for additional genetic material beyond the genetically marked strains already held by several of the service collections. Thus, many collections are building up collections of plasmid-bearing strains or genetically manipulated strains, and are starting the storage of native DNA so that the requirements of geneticists can be met in the future. Development into this area requires the establishment of protocols for the preservation of such material in an unchanged condition, the procedures necessary for quality control purposes and distribution systems. Many service collections have on-going research projects in this area and have appointed geneticists to their staff.

There are rapid developments in the area of computer services and the most challenging of these relate to on-line access to databases. The use of computers for collection, administration and catalogue production is becoming standard in the major collections, and on-line services for registered users for ordering, electronic mail, catalogue searching, and even computer-assisted identification (Chapter 5.2.5) are being developed. The use of electronic mail by culture collections using international computer networks is an important innovation, since it establishes close links not only between the centres and their users, but also between the users themselves and can be expanded to set up computer conferencing between microbiologists with common interests, using the service collections or microbial databases as co-ordinating centres.

The role of service collections as training and research centres is fundamental to the development of microbiology. No other laboratories have the absolute requirement and therefore the commitment to the taxonomy, identification and preservation of microorganisms, and the research carried out by scientists in the collections leads to on-going development in these essential areas. The specialist skills developed as a result of both routine work and research in the collections is transferred to others through participation in training courses throughout the world.

1.5 Safeguarding resource centres for the future

Many service collections have been built up gradually over a long period of time and represent a substantial investment in terms of time and effort. It is important therefore that the investment should not be lost and that the microorganisms collected are conserved for posterity. The isolation, characterisation and preservation techniques used in the collections require specialised knowledge, and the services offered have been developed through long experience; it follows that these units of unique expertise should not be lightly disbanded. Where resources for their support become limited, every effort should be made to make satisfactory alternative provision for the safeguarding of the cultures and transfer of specialised technology.

More commonly it is the smaller specialist collections that become endangered, either through changes in direction of research or through the retirement or transfer of the scientist responsible for the establishment of the collection. In these circumstances the curator should consider the future safety of the collection in good time to allow the necessary transfer arrangements to be made. If a suitable service collection is known, approaches should be made to establish the willingness of the collection to take on additional strains. It is the policy of most major collections to accept cultures that are endangered, although there may be logistical problems in accepting either large numbers or fastidious strains.

The World Federation for Culture Collections has established an Endangered Collections' Committee which may be approached by any microbiologist or organisation facing difficulties with the continuing maintenance of collections. The committee will consider requests for help and has funds, provided by the United Nations Environment Programme (UNEP) and the International Union of Microbiological Societies, to enable experts to visit endangered collections, help with any necessary documentation, identify a suitable recipient, and help with transfer costs and possible short-term emergency preservation. It is the policy of the committee to place endangered cultures in collections that reach certain minimum standards and that are in areas that would be scientifically enhanced by the availability of new microbiological material of the kind endangered. No transfer is proposed without full consultation with experts and with the people concerned with the existing and future collections, nor without the agreement of all parties concerned. The committee is there to respond to requests for help and to do everything possible to ensure the continued existence and availability of unique genetic material.

1.6   List of service resource centres

This section lists the culture collections which are recorded for the most part in the WFCC's World Data Center *World Directory of Collections of Microorganisms* (see Section 1.2) as having a significant collection of yeasts either in terms of numbers, scientific interest or regional importance. The list is not exhaustive and will be amended and updated in future editions as new collections become established or develop as service centres.

Each entry provides the name, acronym, address, telephone, telex and electronic mail numbers and, in most cases, a short statement on the history, holdings and special interests and activities of the collection. Acronyms are listed in Table 1.1, and ordering procedures and services are summarised in Tables 1.2–1.4. It should be remembered that new developments are taking place all the time and that the lists and tables may become out of date quite quickly; collections should be contacted for up-to-date information. Charges for cultures and services are not quoted, since these change frequently, and contact should be made directly with the collection for this information.

The collections are listed according to the geographical area in which they are located (Asia, Australasia, Europe, North America, South America).

1.6.1   *Asia*

Culture Collection and Research Center   CCRC
Food Industry Research and Development Institute
PO Box 246
Hsinchu 30038
Taiwan
China
Tel.: (035) 223191
Holding: Yeasts 670
Total 3000 (bacteria including actinomycetes, filamentous fungi, yeasts, DNA vectors and hosts)   *contd. p. 17 . . .*

Table 1.1. *Culture collections and their acronyms (see individual entries in Section 1.6 of this chapter for full addresses)*

Asia

CCRC	Culture Collection and Research Center, Taiwan, China
CGMCC	Centre for General Microbiological Culture Collections, Beijing, China
IAM	Institute of Applied Microbiology, University of Tokyo, Japan
IFO	Institute for Fermentation, Osaka, Japan
JCM	Japan Collection of Microorganisms, Saitama, Japan
KUKENS	Centre for Culture Collections of Microorganisms, Istanbul, Turkey
NCIM	National Collection of Industrial Microorganisms, Pune, India
TISTR	Thailand Institute of Scientific and Technological Research, Bangkok, Thailand

Australasia

FRR	CSIRO Division of Food Research, North Ryde, Australia

Europe

CBS	Centraalbureau voor Schimmelcultures, Yeast Division, Delft, Netherlands
CCUG	Culture Collection of the University of Göteborg, Sweden
CCY	Czechoslovak Collection of Yeasts, Bratislava, Czechoslovakia
CECT	Coleccion Espanola de Cultivos Tipo, Valencia, Spain
DBV-PG	Collection of the Dipartimento di Biologia Vegetale, Perugia, Italy
DSM	Deutsche Sammlung von Mikroorganismen, Braunschweig, Federal Republic of Germany
LWG	Bayerische Landesanstalt für Weinbau und Gartenbau, Würzburg, West Germany
NCAIM	National Collection of Agricultural and Industrial Microorganisms, Budapest, Hungary
NCPF	National Collection of Pathogenic Fungi, London, UK
NCYC	National Collection of Yeast Cultures, Norwich, UK
RIVE	Research Institute for Viticulture and Enology, Bratislava, Czechoslovakia
VKM BKM	All-Union Collection of Microorganisms, Moscow Region, Pushchino, USSR
VTT	VTT Collection of Industrial Microorganisms, Espoo, Finland

North America

ATCC	American Type Culture Collection, Maryland, USA
NRRL	Agricultural Research Service Culture Collection, Peoria, USA
YGSC	Yeast Genetic Stock Center, Berkeley, California, USA

South America

ESALQ	Departamento Tecnologia Rural, Sao Paulo, Brasil

Table 1.2. Procedures for ordering cultures[a]

Collection	Cash with order (CWO) Cash against invoice (CAI)	Cheques made out to	Currency accepted[b]	Usual delivery time	Form in which orders accepted
Asia					
CCRC	CWO, CAI	Food Industry Research & Development Institute	US dollars NT dollars money orders	5 days	letter, telephone
CGMCC	CWO	Institute of Microbiology Academia Sinica	US dollars, RMB Yuan	2–3 weeks	letter
IAM	–	–	–	1 week	letter
IFO	CAI	IFO	Japanese yen	5 days	letter
JCM	CWO	RIKEN	any major currency	A few weeks	letter, cable, telex
NCIM	CWO, CAI	National Chemical Laboratory, Pune	US dollars, pounds sterling, Indian rupees	2–3 days	letter
TISTR	CWO, CAI	TISTR Culture Collection	baht, US dollars	1–2 weeks	letter, telex, telephone
Australasia					
FRR	CWO, CAI	Collector of Public Resources	US dollars, Australian dollars, pounds sterling	within 3 weeks	letter

Europe					
CBS	CWO, CAI	CBS	any	1–2 weeks	letter, telephone, telex
CCUG	CWO, CAI	CCUG	any	–	letter, telephone, electronic mail
CCY	CWO, CAI	Institute of Chemistry SA Sci, Bratislava	convertible	1–3 weeks	letter, telephone
CECT	CAI	CECT		2–3 days	letter, telephone
DBV-PG	CWO	Dipartimento di Biologia Vegetale, Perugia	–	1–2 weeks	letter, telex, telephone
DSM	CWO, CAI	Deutsche Sammlung von Mikroorganismen	any major currency	same day (lyophilised), 3–7 days other	letter, telex telephone
KUKENS	–	–	–	1–2 weeks	letter
LWG	CAI	Bayerische Landesanstalt für Weinbau und Gartenbau	Deutschmark	2–4 weeks	letter
NCAIM	CWO	Hungarian National Bank	forint, pounds sterling, US dollars, Deutschmark, Swiss franc	3–14 days	letter, telex, telephone

Table 1.2. (cont.)

Collection	Cash with order (CWO) Cash against invoice (CAI)	Cheques made out to	Currency accepted[b]	Usual delivery time	Form in which orders accepted
NCPF	CAI, CWO	Public Health Laboratory Service	pounds sterling	1–4 weeks	letter, telex, telephone
NCYC	CAI, CWO	Institute of Food Research	any major currency	same day (lyophilised), 3 days other	letter, telex, telephone, electronic mail
RIVE	CWO, CAI (free for teaching purposes)	RIVE, through Czechoslovak Commercial Bank		2–3 weeks	letter
VKM BKM	cultures supplied free in USSR, and on an exchange basis elsewhere				
VTT	CAI, CWO	VTT, Biotechnical Laboratory	Finmark, Deutschmark, US dollars, pounds sterling	within 1 week	letter, telex, telephone
North America ATCC	CAI, CWO	ATCC	any major currency	2 weeks	letter, telex, telephone

NRRL	CAI, CWO charges for patent strains only	Agricultural Research Service US Department of Agriculture	US dollars (cheques only)	1 week	letter, telex, electronic mail
YGSC	CAI, CWO	Regents of the University of California	US dollars	2–3 weeks	letter, telephone
South America ESALQ	CAI, CWO	Departamento Tecnologia Rural	US dollars, cruzados	3 weeks	letter, telex, telephone

[a] This table is subject to change and provides a general indication only.
[b] Many collections accept UNESCO coupons.

Table 1.3. *Information services*[a] (see Chapter 2)

Collection	Catalogue produced	Date last edition	General information				Computer searches of databases[b, c]		
			letter	telex	telephone	electronic mail	in house	on-line to outside users	other databases
Asia									
CCRC	+	1986	+	–	+	–	+	–	–
CGMCC	+	1982	+	–	–	–	–	–	–
IAM	+	1979 (JFCC Catalogue)							
IFO	+	1984	+	–	+	–	–	–	–
JCM	+	1986	+	+	+	+	+	–	–
NCIM	+	1977	+	–	+	+	+	–	–
TISTR	+	1985	+	–	+	–	+	–	–
Australasia									
FRR	–		–		–	–	–	–	–
Europe									
CBS	+	1983	+	+	+	(+)	(+)	(+)	(MINE)
CCUG	+	1986 continuously up-dated: floppy discs or paper	+	–	+	+	+	(–)	(MSDN)
CCY	+	1986	+	–	+	(–)	+	–	–
CECT	+	1985	+	–	+	–	–	–	(MSDN)
DBV-PG	+	1965	+	+	+	–	+	–	–
DSM	+	1984	+	+	+	–	+	–	(MINE)
KUKENS	+	1982	+	–	+	–	–	–	(MSDN)
LWG	–	–	+	–	–	–	+	–	–

Collection		Year						Computer services[c]
NCAIM	+	1984	+	+	−	+	−	(MSDN)
NCPF	+	1967	+	−	+	−	+	MiCIS (MINE) / MSDN via MiCIS
NCYC	+	1984 continuously up-dated on-line	+	+	+	+	+	MiCIS (MINE) / COMPASS, MSDN via MiCIS
RIVE	+	1986	+	−	+	−	+	−
VKM BKM	+	1981	information services at present only within USSR					−
VTT	+	1985	+	+	(−)	+	−	−
North America								
ATCC	+	1986 annual up-date	+	+	−	+	+	MSDN
NRRL	−	–	+	−	+	+	−	−
YGSC	+	1984 biennial up-date	+	−	+	−	+	−
South America								
ESALQ	+	1986	−	−	−	−	−	−

( ), planned.

[a] This table is subject to change and collections should be contacted directly for up-to-date information.

[b] All collections listed are affiliated to the WFCC's World Data Center on Collections of Microorganisms.

[c] MiCIS, Microbial Culture Information Service; MINE, Microbial Information Network in Europe; COMPASS, NCYC's on-line computer services; MSDN, Microbial Strain Data Network.

Table 1.4. *Services other than culture supply and information*[a]

Collection	Identification	Preservation	Patent depositary (BT)[b]	Safe-deposit	Contract work by negotiation
Asia					
CCRC	+	+	−	+	−
CGMCC	+	+	+	−	−
IAM	−	−	−	−	−
IFO	−	+	+	+	+
JCM	+[c]	+	−	−	+
NCIM	−	+	+	+	−
TISTR	−	+	−	+	+
Australasia					
FRR	−	−	−	−	−
Europe					
CBS	+	+	+(BT)	+	+
CCUG	+	+	−	+	+
CCY	+	+	+	+	+
CECT	+	+	+	−	−
DBV-PG	+[d]	+	−	−	−
DSM	+	+	+(BT)	+	+
KUKENS	+	+	−	−	−
LWG	−	−	−	−	−
NCAIM	+	+	+	+	+
NCPF	+	−	−	−	−
NCYC	+	+	+(BT)	+	+
RIVE	+	+	−	−	−
VKM BKM	services at present available only within USSR				
VTT	+	+	−	+	+
North America					
ATCC	+	+	+(BT)	+	+
NRRL	+	−	+(BT)	−	−
YGSC	−	+	−	−	−
South America					
ESALQ	−	−	−	−	−

[a] This table is subject to change and collections should be contacted directly for up-to-date information.
[b] International Depository Authorities under the Budapest Treaty.
[c] By negotiation.
[d] DNA–DNA reassociation studies carried out.

1.6.1 *Asia* (*contd. from p. 8*)

Centre for Microbiological Culture Collection   CGMCC
China Committee for Culture Collections of Microorganisms
(CCCCM)
Institute for Microbiology
PO Box 2714
Academia Sinica
Beijing
China
Tel.: 285614
Holding: Yeasts 1450
Total 7690 (bacteria, actinomycetes, yeasts, filamentous fungi,
basidiomycetes)

The CGMCC is a member of CCCCM as well as being the Department of
Culture Collections of the Institute of Microbiology in the Chinese
Academy of Sciences. The collection maintains bacteria, actinomycetes,
yeasts and filamentous fungus cultures used for taxonomic and biochemical research, education, biosynthesis of organic compounds,
physiological assay, biological control and traditional fermentation.

The CGMCC is engaged in the collection, identification, preservation,
research and distribution of cultures of microorganisms with significance to industry, agriculture and pharmaceutics. In addition, the
centre is engaged in work on cryobiology, using ultra-low temperature
freezing for the preservation of cultures and psychrophiles. A data bank
on cultures of microorganism is planned to be established at CGMCC.

In April 1985, the CGMCC was entrusted by the China Patent Office
to form the repository with responsibility for accepting deposits of
strains of microorganisms for patent applications. After the patent is
granted, the strains connected with the patent procedures will be
published in the national catalogue of cultures.

Institute of Applied Microbiology   IAM
The University of Tokyo
1-1-1 Yayoi, Bunkyo-ku
Tokyo 113
Japan
Tel.: 03 812 2111
Holding: Yeasts 400
Total 2600 (bacteria, filamentous fungi, yeasts, algae)

The IAM Culture Collection was established in 1953 to promote research in basic and applied microbiology by collecting, preserving and distributing authentic cultures. This is one of the main purposes for which the Institute of Applied Microbiology was founded.

The IAM Culture Collection is a member of the Japan Federation for Culture Collections.

Cultures of bacteria, yeasts, filamentous fungi and micro-algae are maintained, and cultures are distributed to institutions both in Japan and abroad. Cultures are preserved by freeze-drying, L-drying (drying from liquid state *in vacuo*), cryogenic freezing and serial transfer. Deposition of cultures of scientific interest is welcomed.

Institute for Fermentation, Osaka   IFO
17–85 Juso-honmachi 2-chome
Yodogawa-Ku
Osaka 532
Japan
Tel.: 06 302 7281
Holding: Yeasts 2330
Total 12 400 (bacteria, filamentous fungi, yeasts, phage, animal cell lines)

The Institute for Fermentation, Osaka, is the largest general collection of microorganisms in the Japan Federation for Culture Collections. It contains bacteria, actinomycetes, moulds, yeasts and animal cell lines. The yeast collection maintains about 2330 cultures, representing a wide range of yeasts, including type strains, reference strains, historically significant strains, strains from a variety of sources, strains for bioassay, biochemical and genetic research, industry and education. The IFO also maintains genetically marked strains, recombinants and plasmid-bearing strains of importance to research and technology.

All yeast cultures are preserved by L-drying (drying from liquid state *in vacuo*). The L-drying has proved suitable for long-term maintenance of strains carrying recombinant plasmids. Some strains are kept frozen at −80 °C for additional security.

The IFO is a recognised depository authority in patent applications of the European Patent Organisation, the German Patent office and patent authorities of other countries. The collection also accepts cultures for safe-deposit.

The IFO is pleased to answer general enquiries on the selection of suitable strains for particular purposes and to provide information relating to individual strains.

Japan Collection of Microorganisms   JCM
RIKEN
Wako-shi
Saitama 351-01
Japan
Tel.: 0484 62 1111
Telex 2962818 (RIKENJ)
Cable RIKAGAKUINST
Holding: Yeasts 1400
Total 5000 (bacteria, actinomycetes, filamentous fungi, yeasts)

The Japan Collection of Microorganisms (JCM) was founded in 1980 as part of the national policy for the promotion of life sciences, and it is expected to function as a major microbiological center in Japan. It collects, preserves and distributes bacteria (including actinomycetes), yeasts and moulds of scientific, industrial, medical and ecological importance, in active co-operation with other collections. Also, since it shares the same institute (RIKEN) as the World Data Center on Collections of Microorganisms, JCM will contribute to the activities of the WDC. RIKEN co-operates with the Japan International Cooperation agency in training young life-scientists from developing countries and is contracted to the Chinese Academy of Sciences for the exchange of research workers.

National Collection of Industrial Microorganisms   NCIM
Division of Biochemical Sciences
National Chemical Laboratory
Pune 411008
India
Tel.: 56451, 52, 53
Telex 0145.266
Holding: Yeasts 800
Total 3000 (bacteria, filamentous fungi, yeasts protozoa)

The National Collection of Industrial Microorganisms maintains over 3000 cultures of microorganisms having applications in basic and applied microbiology and biotechnology. It has been in existence at the

National Chemical Laboratory, Pune for over 35 years and has been primarily responsible for the supply of authentic strains of microorganisms to industry as well as academic institutions. It maintains close links with the other collections through membership of the WFCC and the SE Asian MIRCEN. NCIM maintains strains of importance to biotechnology and recently has been actively involved in the screening of rare or little-investigated microbial systems from indigenous resources and investigating their potential for biotechnological applications. For example, studies have been made on the isolation of *Conidiobolus* strains having high activity of alkaline proteinase and *Chainia* species producing extracellular substrate specific glucose and xylose isomerase and high activity xylanase. Cultures used in genetic engineering and recombinant DNA technology are included in the collection.

Thailand Institute of Scientific and Technological Research
TISTR

196 Phahonyothin Road
Bangkok 10900
Thailand
Tel.: (02) 5791121
Telex 21392 TISTR TH
Holding: Yeasts 250
Total 2000 (bacteria, filamentous fungi, yeasts, algae)

The TISTR Culture Collection originated from a collection of fungi which was isolated from natural habitats of tropical dry-evergreen forests in Thailand, under a research project on biodeterioration carried out by TISTR during 1967–8. The Collection has been gradually developed to include collections of non-pathogenic bacteria and fungi, including yeasts of industrial and educational interest. Most strains of microorganisms are preserved by lyophilization with a minority under liquid nitrogen.

In 1976, TISTR Culture Collection was appointed by UNESCO to serve as a Microbiological Resource Center for Southeast Asia (Bangkok MIRCEN). Activities of the MIRCEN at Bangkok include development of culture collections in the region, with emphasis on:

(1) exchange of economically important microbial strains;
(2) training of culture collection personnel;
(3) fellowship programme; and
(4) promotion of research on organisms in areas of microbiology

appropriate to Southeast Asia, including inter-regional co-operation on specific projects.

The Bangkok MIRCEN serves the scientific community of Southeast Asia through its network of co-operating laboratories in Hong Kong, Indonesia, Malaysia, Phillipines, Singapore and Thailand.

1.6.2   *Australasia*

CSIRO Division of Food Research   FRR
PO Box 52
North Ryde
New South Wales 2113
Australia
Tel.: (612) 8878333
Telex AA 23407
Holding: Yeasts 100
Total 3400 (filamentous fungi, yeasts)

The FRR culture collection consists mostly of industrially important fungi, particularly those of significance in food spoilage and mycotoxin formation, although yeasts are also held. Both moulds and yeasts are kept in lyophil storage.

The collection contains a very large number of isolates from Australasian sources not held in other collections, and these represent an untapped source of physiologically and biochemically significant organisms.

1.6.3   *Europe*

Centraalbureau voor Schimmelcultures   CBS
(Yeast Division)
Julianalaan 67
2628 BC Delft
Netherlands
Tel.: 015-782394
Telex 38151 BUTUD NL
Holding: Yeasts 4500

The yeasts of the CBS collection are housed in the Laboratory for Microbiology at the Technological University of Delft. The collection holds a diverse selection of yeasts of taxonomic as well as practical and research importance such as assay and industrial strains. The collection also accepts strains of bacteria as well as yeasts in connection with patent applications, and is a recognized depository under the Budapest Treaty. Cultures are preserved by freeze-drying and by freezing in liquid nitrogen. The cultures are continually being checked, and any cultures that are ordered are checked before they are sent.

The CBS offers an identification service for a fixed fee for each strain and can also accept some contract projects for a negotiated fee. In addition, instruction in identifying and classifying yeasts is given. Enquiries about contract work and patent deposits should be made to the head institute (PO Box 273, 3740 AG Baarn).

The collection issues a List of Cultures in book form and on microfiche; more detailed catalogue information will be available on-line at the main institute through the MINE network (see Chapter 2). A (free) newsletter is issued in the spring and autumn of each year.

Research is done into the systematics of yeasts, and the research staff are usually willing to co-operate with outside workers in the examination of putative new species and preparation of descriptions of taxa.

Culture Collection of the University of Göteborg   CCUG
Department of Clinical Bacteriology
Guldhedsgatan 10
S 41346 Göteborg
Sweden
Tel.: 46.31 602016
(Electronic Mail TIXCC AT SEGUC 21, DIALCOM 42:
CDT0070, TELECOM GOLD 75:DBI0070)
Holding: Yeasts 300
Total 20 000 (bacteria, filamentous fungi, yeasts)

The CCUG holds and despatches a large number of lyophilized cultures as well as giving general information or information pertaining to its own strains. As it is well equipped for freeze-drying, it plans to increase the yeast part of the collection to help scientists in Scandinavia in areas of biotechnology and clinical microbiology. Since it is working with microcomputers, new databases and files may be created by its own staff whenever the need arises. Orders, confirmation of orders, database information or information retrieval for customers in external databases may be performed exclusively by electronic systems. The CCUG catalogue of strains (300 pages, or 750 kb, at November 1986) is continuously updated and available as printouts or on floppy disks.

Czechoslovak Collection of Yeasts   CCY
Institute of Chemistry
Slovak Academy of Sciences
Dúbravská cesta 9
842 38 Bratislava
Czechoslovakia
Tel.: 375000
Holding: Yeasts, yeast-like organisms, heterotrophic
algae 4000

The Czechoslovak Collection of Yeasts was established by Dr A. Kocková-Kratochvilová in Prague in 1942. Since its foundation the CCY has become established in the Institute of Chemistry of the Slovak Academy of Sciences. Recently the collection has become increasingly important, since biotechnological research is a fundamental part of CCY's activities. The collection maintains about 4000 yeasts, yeast-like organisms and heterotrophic algae. Strains are preserved under paraffin oil, by freeze-drying and in liquid nitrogen. Yeast strains are regularly evaluated morphologically and physiologically. About 100 features are coded for storage in a computer, and on the basis of this coding a computer program has been established for identification of yeasts. The publication of the codes belonging to individual collection strains in the Catalogue of Yeast Cultures serves as a databank for users.

The CCY collection collaborates closely with six other Czechoslovak collections of yeasts in Prague and Bratislava, which maintain specialist industrial and genetical strains.

The CCY collection supplies 150 Czechoslovak institutions with yeast strains. Charges are made for strains and services. Many strains are also sent to other parts of the world on an exchange basis. Services and cultures are provided for pure or applied research, for industrial production, for teaching in universities, for nutritional, fermentation, agricultural, pharmacological or similar industries.

CCY collaborates with the Czechoslovak Patent Office as a depository for patent strains in Czechoslovakia and, in the future, also for international deposits.

Colección Espanola de Cultivos Tipo   CECT
Departamento de Microbiologia Facultad de Ciencias
Biologicas
Burjasot

Valencia
Spain
Tel.: 96 3630011
Holding: Yeasts 160
Total 900 (bacteria, filamentous fungi, yeasts)

The Colección Espanola de Cultivos Tipo maintains 900 microbial strains, about 500 of which are bacteria and 400 fungi (including yeasts). It supplies strains that are ordered by telephone or letter, and a charge is made to industrial users.

The CECT accepts deposits of strains free of charge and also accepts deposits for national patent purposes, for which a charge is made. It can provide information about certain aspects of microbiology and some identification services are available.

Some of the strains preserved in the CECT are of biotechnological interest. Work associated with the CECT is carried out on wine yeasts and lactic bacteria, and also on transformation in filamentous fungi, all of which have importance for biotechnology.

Yeast Collection of the Dipartimento di Biologia   DBV-PG Vegetale of the University of Perugia

c/o Dipartimento di Biologia Vegetale
Via 20 Giugno, 74
I-06100 Perugia
Italy
Tel.: 39 75 31766
Telex 216842 UMPG 11
Holding: Yeasts and yeast-like fungi 3000

The Yeast Collection DBV-PG was started in the early 1920s with the main purpose of conserving the most taxonomically and biotechnologically significant cultures isolated and classified in the course of a large survey on the ecology of yeasts in nature, conducted by the Institute of Agricultural Microbiology of the University of Perugia, Italy. An additional objective is the development of a reservoir for molecular taxonomy studies based on the comparison of informational macromolecules.

An atypical aspect that makes the DBV-PG collection unique is the fact that a major portion of the cultures belongs to a handful of wine- or must-associated species, each represented by a large number of strains.

For example, *Saccharomyces cerevisiae* is represented by more than 1000 isolates.

Section A of the collection consists of yeasts associated with the fermentation industry and has recently been revised taxonomically and technologically. Section B includes about 400 cultures from other collections including type strains. Section C includes species isolated from various natural habitats not associated with the fermentation industry. All of the cultures in this section need to be reclassified according to the latest monograph on yeast taxonomy.

Deutsche Sammlung von Mikroorganismen DSM
Mascheroder Weg 1b
D-3300 Braunschweig.
Federal Republic of Germany
Tel.: 0531-6187-0
Telex 9 52 667 GEBIO-D (Electronic mail TELECOM GOLD 75:DBI0178)
Holding: Yeasts 250
Total *c.* 5000 (bacteria, filamentous fungi, yeasts)

The DSM, which is part of the Institute for Biotechnological Research (GBF), is continually increasing in size, and many new strains of applied or biotechnological interest are accessioned each year. As well as acting as a depository and source of cultures, the DSM offers a number of scientific services. These include identification, preservation, safe-deposit, training and consultation services. In addition the DSM is an International Depositary Authority under the Budapest Treaty. The DSM is registered with the World Federation for Culture Collections' World Data Center and is a member of the WFCC and ECCO (see Chapter 8).

Center for Culture Collections and Microorganisms KUKENS
Istanbul Tip Fakultesi
Mikrobiyoloji Anabilim Dali
Temel Bilimler Binasi
Capa 34390
Istanbul, Turkey
Tel.: 5250904-5255504

Holding: Yeasts 150

Total 1440 (bacteria, filamentous fungi, yeasts, viruses, cell cultures, protozoa)

The Center for Culture Collections of Microorganisms (KUKENS) was established in 1979 and is associated with the Istanbul Faculty of Medicine. KUKENS is a member both of the WFCC and the European Culture Collections' Organization. The collection is also registered with the World Data Center for Collections of Microorganisms.

The main function of KUKENS is to keep medically important microorganisms used in the manufacture of industrial products and natural fertilisers, in pilot investigations in medicine and industrial microbiology, in investigations in the fields of human and veterinary medicine, biochemistry, agriculture, pharmacy, molecular biology, cytology and genetics.

In addition to its culture collection activities, KUKENS acts as a training centre.

Bayerische Landesanstalt für Weinbau und Gartenbau   LWG

Residenzplatz 3
D-8700 Würzburg
Federal Republic of Germany
Tel.: 0931/50701
Holding: Yeasts 900

The LWG collection was established in 1959 mainly with isolates from grapes, grape juice and wine. The collection maintains about 900 yeasts and related yeast species, for the most part those of oenological interest. It also includes many killer strains.

Charges are made for cultures supplied for industrial purposes. Strains are also supplied on an exchange basis with other collections.

National Collection of Agricultural and Industrial Microorganisms   NCAIM

H-1118 Budapest
Somloi ut 14-16
Hungary
Tel.: 00 36 1 665466
Telex 226011
Holding: Bacteria, filamentous fungi, yeasts 1500

The National Collection of Agricultural and Industrial Microorganisms (NCAIM) was established in 1974 at the Department of Microbiology, University of Horticulture in Budapest, Hungary, and is the responsibility of the Ministry of Agriculture and Food. The collection contains the strains of about 30 research and higher education institutes engaged in agricultural and industrial activity. Bacteria, filamentous fungi and yeasts, among them a special wine yeast collection, are held.

All strains of microorganisms are preserved by freeze-drying and freezing in liquid nitrogen. In addition, traditional preservation methods are used in parallel.

The NCAIM acquired the status of an International Depositary Authority for patent cultures on June 1, 1986, and accepts strains for deposit in connection with patent applications under the terms of the Budapest Treaty.

Data on the strains maintained are being computerised.

National Collection for Pathogenic Fungi   NCPF
Mycological Reference Laboratory
Central Public Health Laboratory
61 Colindale Avenue
London NW9 5HT
United Kingdom
Tel.: 01 200 4000
Telex 8953942 DEFEND (G)
Holding: Yeasts 350
Total 1000 (filamentous fungi, yeasts)

The National Collection of Pathogenic Fungi maintains yeasts, moulds and aerobic actinomycetes isolated from infections of man and other warm-blooded animals. It forms part of the UK network of microbiological culture collections, and was once housed at the London School of Hygiene and Tropical medicine (NCPF strain numbers have in the past borne other acronyms such as LSH or MRL).

The collection covers a representative number of pathogenic *Candida* and *Cryptococcus* species, mostly from UK infections but including some standard strains of international importance which have been the subject of much experimentation.

National Collection of Yeast Cultures   NCYC
Institute of Food Research, Norwich Laboratory
Colney Lane
Norwich
Norfolk NR4 7UA
United Kingdom
Tel.: (0603) 56122
Telex 975453 (FRINOR G)
(Electronic Mail TELECOM GOLD 75:DBI0013/DBI0151 0005
            COMPASS: on registration with NCYC
            Computer Services (see below))
Holding: Yeasts 2300

The National Collection of Yeast Cultures is the part of the UK network of service collections responsible for yeasts other than known pathogens. Until 1947 it formed part of the National Collection of Type Cultures, London, and following a period at the Brewing Research Foundation, Nutfield, has been housed at the Agricultural and Food Research Institute's Norwich laboratory.

The collection holds representatives of a wide range of genera and species, but has specialist sub-collections relating to industrial and genetically important strains. The industrial collection contains over 350 strains of *Saccharomyces cerevisiae*, most of which have been characterised for their industrial performance in fermenters. The genetic collection contains strains of *Saccharomyces cerevisiae* and *Schizosaccharomyces pombe* with defined nutritional and resistance markers, mating types and other genetic characteristics. All strains are maintained both by cryopreservation in liquid nitrogen at −196 °C and lyophilization. On-going routine checks are made for purity and authenticity.

On-line computer services for ordering cultures, catalogue viewing and searching, electronic mail, probabilistic yeast identification (see Chapter 5) and NCYC news board are available for users registered with the Collection. The Collection has all catalogue and strain data computerised and available for searching on MiCIS (Microbiol Culture Information Service), and there are plans to include some of this data in the MINE database (see Chapter 2). Further information on computer services may be obtained from the collection.

The NCYC is a member of the World Federation for Culture Collections, the European Culture Collections' Organization and the UK Federation for Culture Collections. It is an International Depositary Authority for patent deposits under the Budapest Treaty.

The collection has an on going research programme, collaborates with national and international training programmes and offers individual training for specific purposes.

Research Institute for Viticulture and Enology   RIVE
Matuskova 25
83311 Bratislava
Czechoslovakia
Tel.: Bratislava 46224/25/26
Holding: Yeasts 800

RIVE provides authentic strains of yeast for industry, research and teaching purposes to universities in Czechoslovakia and abroad. 'Liquid' pure wine yeasts for the Czechoslovak wine industry are produced and distributed, as are yeasts for active dry wine yeast production in Czechoslovakia. Yeasts are exchanged with overseas yeast collections. Identification of yeasts for Czechoslovak industrial and teaching institutions is carried out. Training is given to university students and post-graduate students. Technical and scientific information and assistance in taxonomic, ecological and biochemical research is provided.

All-Union Collection of Microorganisms   VKM BKM
Institute of Biochemistry and Physiology of Microorganisms
Moscow Region
Pushchino 142292
USSR
Tel.: 3 05 26
Telex 3205887 OKA
Holding: Yeasts 2000
Total 8000 (actinomycetes, bacteria, filamentous fungi, yeasts)

The VKM BKM collection maintains over 2000 cultures of yeasts covering almost all genera and species, except those pathogenic to man. Many of the yeasts are type strains. Deposited strains are checked morphologically and physiologically. Cultures, which are maintained mostly by freeze-drying, are supplied free of charge to research, industrial and educational laboratories within the USSR. Cultures are supplied to other countries mostly on an exchange basis. The interests of the VKM lie chiefly in taxonomy, identification and ecology. As well as the publication of a catalogue of cultures, a list of new accessions is published from time to time in Prikladnaya Biochimiya Mikrobiologiya.

VTT Collection of Industrial Microorganisms   VTT
VTT Biotechnical Laboratory
Tietotie 2
SF-02150 Espoo
Finland
Tel.: 358 0 45615133 or 358 0 4565133
Telex 122972 VTTHA SF
(Electronic Mail: COMPASS)
Holding: Yeasts 300
Total 900 (bacteria, yeasts, filamentous fungi)

The VTT collection is a national service collection in Finland. The first strains were deposited in 1959. The present holding is about 900 strains, including approximately equal numbers of yeasts, filamentous fungi and bacteria. More than half of the yeast collection consists of *Saccharomyces cerevisiae* strains, including bakers' yeasts (48), brewers' yeasts (75), distillers' yeasts (5), wine yeasts (14), sour dough yeasts (7) and other strains (23). The collection is increasing at a rate of about 10% per year.

The main functions of the collection are the deposition (but not for patent purposes) and preservation of cultures, and the delivery of pure cultures to industry, research and education in Finland. The VTT cultures are maintained mainly in the freeze-dried state and at ultra-low temperatures.

The VTT collection was registered with the Nordic Register of Microbiological Culture Collections (NORCC) in 1981 and with the World Data Center for Microorganisms (WDC) in 1986. The collection has been a member of the European Culture Collections' Organization (ECCO) since 1984 and of the World Federation for Culture Collections (WFCC) since 1985. The collection is financed by the Technical Research Centre of Finland (VTT).

1.6.4 *North America*

American Type Culture Collection   ATCC
12301 Parklawn Drive
Rockville
Maryland 20852
USA
Tel.: (301) 881 2600
Telex 908 768 ATCC NORTH
(Electronic Mail: DIALCOM 42: CDT0004)
FAX: 301 231 5826
Holding: Yeasts 8000
Total 40 000 (most microbiological groups, plus animal and plant tissue cultures)

The American Type Culture Collection (ATCC) is a non-profit organisation which houses one of the most diverse collections of strains in the world, with over 40 000 different strains of animal viruses, bacteria, bacteriophages, cell lines, fungi, plant viruses, protists and yeasts, plus rapidly growing collections of oncogenes, recombinant DNA vectors, hybridomas, plant tissue cultures, and human DNA probes and libraries.

Research at the ATCC includes comparative microbiology, microbial systematics, computer-assisted identification analyses, and improved methods for the isolation, propagation, characterisation and preservation of strains, strain improvements for industrial applications, eukaryotic gene regulation, studies on the mechanism of action of oncogenes and artificial cultivation of exotic mushrooms.

Microbiological Research Services offers culture identification, cryopreservation, freeze-drying of microorganisms and biological reagents, culture safekeeping, plasmid preparation, mycoplasma testing, karyotyping and virus purification.

The ATCC is an approved International Depositary Authority (IDA) under the Budapest Treaty on the International Recognition of the Deposit of Microorganisms for the Purposes of Patent Procedures, and is also an official depository for patent procedures under the European Patent Organization (EPO), the United States, Germany, Japan, and most other countries.

In addition, ATCC's workshop series offers scientists and technicians hands-on laboratory training in areas such as cryopreservation, quality control and assurance, recombinant DNA methodology, computer usage in microbiology and biotechnology patents.

The ATCC is also the control operations facility of an international data bank on cloned cell lines and their immunoreactive products. An international hybridoma data bank (HDB) has been established and is provided as a free service.

Catalogues for each collection maintained are regularly published and are available free of charge, but postage is charged for shipments outside the USA.

The yeast collection currently comprises over 8000 strains, representing type cultures, reference cultures, genetic stocks and cultures with special applications. The uses of these cultures have applications across biology, medicine, industry and agriculture. Genetic stock cultures have become extremely important to the study of gene organisation, regulation of genetic expression, DNA repair, and movable genetic elements of eukaryotic cells.

In order to reduce loss of viability, contamination, variation, mutation or deterioration, all the cultures maintained in the ATCC are freeze-dried and stored at 4 °C and/or frozen in liquid nitrogen.

Agricultural Research Service Culture Collection   NRRL
Northern Regional Research Center
1815 North University Street
Peoria
Illinois 61604
USA
Tel.: (309) 685 4011
(Electronic Mail: TELEMAIL (User designation RL.MPR))
Holding: Yeasts 14 000
Total 77 000 (actinomycetes, bacteria, filamentous fungi, yeasts)

The ARS Culture Collection (NRRL), which maintains 77 000 strains, is the major microbial culture collection of the Agricultural Research Service, US Department of Agriculture. The collection is comprised of strains primarily of agricultural and industrial significance, consequently relatively few human, animal and plant pathogens are included. Overall division of the collection and number of strains maintained is the following: yeasts, 14 200; filamentous fungi, 44 000; bacteria, 10 200; and actinomycetales, 9100. The collection also serves as an International Depositary Authority for patent cultures under the Budapest Treaty.

No catalogue is issued, but requests may be made by asking for particular taxa, or by strain number if known, or for strains with particular uses. No charges are made for cultures distributed from the general collection and requests are consequently limited to 12 strains.

Yeast Genetic Stock Center   YGSC
Department of Biophysics and Medical Physics
University of California
Berkeley
California 94720
USA
Tel.: (415) 642 0815
Holding: Yeasts of genetic interest 850

The Yeast Genetic Stock Center collects, maintains, monitors and disseminates to investigators all over the world, well-characterised, genetically marked strains of the yeast *Saccharomyces cerevisiae*. For ease of dissemination, strains are stored as dried milk-paper replicates at 4 °C; a back-up supply is maintained at −75 °C in 20% glycerol.

At intervals of two or three years the YGSC publishes and distributes a revised list of available strains describing their genotypes, the source, pertinent references and some general procedures for their use (a voluntary contribution of $10.00 is suggested for recipients of the catalogue). Periodically, strains are checked for markers and ploidy and returned to the collection. When necessary, new strains are generated at the YGSC to replace occasionally diploidized cultures or revertants. New constructs are also developed to facilitate the use of the collection.

Strains of general interest and usefulness that have been well charac-terised are solicited from, or directly deposited by, various investigators; they are included in the collection after having been checked for markers and ploidy.

A fee is charged to users of the stock center to supplement funding by the National Science Foundation. Some training in techniques of yeast genetics is given to graduate and undergraduate students.

The methods of Orthogonal Field Alteration Gel Electrophoresis (OFAGE) and Field Inversion Gel Electrophoresis (FIGE) are being used at the YGSC for the karyotyping of various 'key' strains of *Saccharomyces cerevisiae* and other genera of yeasts as well as for the identification of chromosomal anomalies in some of the strains in the collection, such as translocations and other aneuploidy phenomena.

1.6.5   *South America*

Departamento Tecnologia Rural   ESALQ
Avenue Padua Dias 11
Caixa Postal 9
Piracicaba
San Paulo
Brasil
Tel.: 33 0011 R272
Holding: Yeasts 500
Total 1100 (bacteria, filamentous fungi, yeasts)

Information on the following collections in Brazil has been supplied by the Tropical Data Base (BDT), Campinas (see below). Further information about these and other yeast collections in Brazil may be obtained from the BDT or the culture collections themselves.

Laboratories de Genetica de Microorganisms
Departamento Microbiologia
University of Sao Paulo
Caixa Postal 4365
05508 Sao Paulo, SP
Brasil
Tel.: (011) 210-4311
Holding: Yeasts 500

The collection consists mostly of strains of *Saccharomyces cerevisiae* of genetical interest.

Departamento de Micologia
Universidade Federal de Pernambuco
Rua. Prof. Morais Rego S/N
Cidade Universitaria
50.000 Recife, PE
Brasil
Tel.: (081) 271 3469

An identification service is provided for clinical isolates and isolates from the environment that are principally of agricultural interest.

Laboratorios de Ecologia Microbiana e Taxonomia
Departamento de Microbiologia Geral
Instituto de Microbiologia
Universidade Federal do Rio de Janeiro
Ilha do Fundao/Cidade Universitaria
21944 Rio de Janeiro, RJ
Brasil
Tel.: (021) 590 3093
Holding: Yeasts 200

The collection has a special interest in the taxonomy and ecology of yeasts isolated from tropical soils and fruits and aquatic environments.

Tropical Data Base (BDT) and Tropical Culture Collection (CCT)
Fundacao Tropical de Pesquisas e Tecnologia 'André Tosello'
Rua Latino Coelho, 1301
13085 Campinas, SP
Brasil
Tel.: (0192) 427022
(Electronic Mail DIALCOM 42:CDT0094)
Holding: Yeasts 100

The collection acts primarily as an information and training centre (see Chapter 2). It also has a small collection of yeasts of importance to industry and teaching.

2

Information resources

M. I. KRICHEVSKY, B.-O. FABRICIUS and
H. SUGAWARA

2.1 Introduction

Microbiologists are faced with consideration of exponential growth in their laboratories on a daily basis. As users of a chapter on information resources for biotechnology they are exposed to a double dose of exponential growth. First, the explosion of information technology itself is due to the massive amounts of computing power available at ever diminishing cost. In turn, a population of computer-aware and computer-literate microbiologists present a growing demand for more sophisticated access to modern information technology. The community of information technologists in concert with microbiologists are responding to this demand with a multiplicity of initiatives using various strategies.

The resulting activity induces feelings of inadequacy in the authors of such chapters as this, since at the moment of delivery to the editors the information is out of date. Resources previously known only by rumour are tested. Simple facilities being tested as pilot projects are quickly made available to the community. Local data banks open their doors to regional and even world-wide participation. Databases on databases spring up because of the need to discover available resources. Occasionally, resources fall by the wayside. The net result is an ever increasing base of information resources for biotechnologists.

While the information about information presented in this chapter is out of date as soon as it is written, the resources described are most likely to be improved and be more useful than the descriptions indicate. For information on new developments the listed resources should be contacted.

2.2   Information needs

The need of the biotechnologist for widely disparate categories of information is a consequence of the varied nature of the tasks required to design, develop, and consummate a process. The biotechnologist must find or develop genotypes of the required composition, discover the conditions for expression of the desired phenotypic properties, maintain the clones in a stable form, and describe all of these parameters in a fashion understandable to peers. Most of the categories of information required will be outside the expertise of any one individual. Thus, a panel of experts representing all disciplines involved must be assembled or access to diverse databases must be achieved. The library or publicly accessible databases can lead to the required information sources which range from the traditional scholarly publications to assemblages of factual or primary laboratory observations. This chapter presents an overview of the kinds of information available and the mechanisms to access them. In particular, it will concentrate on information resources for finding material with the desired attributes, be they taxonomic, historical, genetic, or phenotypic.

2.2.1   *Interdisciplinary information sources*

The personal training and professional experience in the particular narrow field in which they work allows microbiologists to perform many daily tasks without reference to outside sources of information. However, the interdisciplinary nature of the practice of biotechnology forces the use of navigational aids to the knowledge base of unfamiliar areas. This point is demonstrated by the observation that approximately 80% of the enquiries to the CODATA/IUIS Hybridoma Data Bank (see below) are from persons who are not immunologists. It follows that a successful database resource should be designed with the interdisciplinary users in mind as they often will form the largest segment of the user population.

The main pathways to locating an existing source of strains with the properties that will be useful in the projected process are through records of the primary observations of properties or through derived information such as taxa or strain designations. In either case, the desired result is one or more strain designations and instructions on where to get cultures. Even though the desired end result is the same with both pathways, the mechanisms for recording and disseminating the information are usually, but not necessarily, quite different.

Culture collections which have a stated mission of providing service to a public user community, especially through distribution of cultures outside of their host institution, tend to use a taxonomic orientation in that their records are commonly kept as discrete strain descriptions, often one strain to a page. The whole strain description is easily read while comparisons among strains are difficult.

Culture collections serving predominantly as local institutional repositories of strains for research, teaching, or voucher specimens for archival storage, tend to rely directly on primary observation data kept in tabular form with the attribute designations as column headings and the strain designations as row labels. In contrast to the previous case, comparisons among strains are easily made, while assembling a complete strain description may require following the row designation for the strain through multiple tables.

Even now, most service collections use traditional paper-based data management methods rather than computers. Within a few years, the majority of collections will be using computers as their main data-handling tool. The functional distinctions between these alternative forms of data organisations blur with the use of computers, but can still be a factor if good information management practices are not followed.

2.2.2   *Where to get strains*

The ultimate source for strains with desired properties is isolation from nature. Indeed, large efforts have been mounted to find strains with desired properties such as production of antibiotics. These efforts are feasible when a good screening procedure, such as zones of clearing around a colony, is available. Even then, the effort is labour-intensive with resulting high costs.

The existence of culture collections with data on the characteristics of the holdings makes possible enormous savings when appropriate strains are available. However, the data must be available to the process developer with reasonable ease. The best source for strains will often be from the collection with the most available data rather than the most complete selection of strains with likely sets of characteristics.

2.2.3   *How to get strains*

The pathways used in obtaining access to the required data for finding desirable strains start with the same foci as the collections themselves. A taxonomic strategy or primary observation strategy may be used. A taxonomic search strategy for strains producing higher concentrations of a particular material (e.g. penicillin, riboflavin, ethanol) might well begin with asking service culture collections for all strains of *Penicillium notatum*, *Ashbya gossypii*, *Saccharomyces cerevisiae* in their collections and screening them for level of production. This method of searching requires the searcher to know which taxa are likely to have the desired attributes.

A primary observation search strategy for organisms with the ability to degrade a particular material might well start with asking for all strains that degraded that material and had the growth characteristics that were desirable under the projected process conditions. The question to the collection might be 'Could you provide me with the characteristics of all your strains able to use hexadecane while growing aerobically at 25 °C?'. This method of searching requires no taxonomic knowledge on the part of the searcher. Clearly, the format of data storage in the collection would markedly affect the relative ease of answering these questions.

2.2.4 Strain data

The traditional classes of data used to describe strains in culture collections still predominate. These include morphological, physiological, biochemical, genetic, and historical data. Clearly, these classes will form the basis for all future collection data as well. The current emphasis on biotechnology places new demands on the informational spectrum desired of microbial and cell line collections. In addition to the requests for taxa with specific properties, information is also requested on potential utility of strains in biotechnological processes and the actual or potential hazards arising from their use. These ancillary but important data are sparse in their availability, but some service collections with a concern for the needs of industry are collecting data of attributes specifically useful in biotechnological processes such as temperature tolerances, behaviour in fermenters, stability of monoclonal antibodies under adverse conditions, toxicity of products, and pathogenicity for unusual hosts.

Physiology. The bulk of information included in publicly accessible databases will be physiological and biochemical. Most queries will be on what the strains may be capable of doing with respect to the desired process, and the databases will be heavily weighted towards containing this kind of information.

While printed compendia of physiological and biochemical data are technically possible, they are rare and, in view of the expense of publishing such volumes, will continue to be rare. Rather, these kinds of data are more reasonably compiled in bulk and disseminated through computers.

Morphology. The detailed morphological data held by most culture collections are generally of less interest than details of physiological data for process development. In various broad classes of organisms (fungi, protozoa, and algae) morphology may be critically important for taxonomy and identification. These same detailed attributes usually have little bearing on the conduct of most processes. However, some basic morphological information will be of fundamental importance in process engineering. For example, cell size knowledge is needed if filtration is a part of the process. Likewise, use of filamentous strains will require more energy for stirring fermenters than non-filamentous forms. Additionally, knowledge of sexual reproduction or fusion attributes are critical if hybridization or strain improvement programmes are to be carried out.

Publicly accessible electronic databases used in searches will, with few exceptions, have only basic or general morphological information on strains because of the costs associated with printing or storage of such information. The decision on how much morphological detail to include in a public database will vary with the interaction between cost and importance of the information.

Industrial. Data on the use of particular strains for specific processes is a frequently requested category of data. Such data exist, but in a widely scattered, uncoordinated fashion. They are contained in the open literature, collection catalogues, patent disclosures, and other more obscure repositories. The same is true of related data on strain behaviour in the processes themselves. There are few biological data compendia equivalent to the materials properties databases available to the engineers in chemistry, metallurgy, and similar disciplines. This situation is partly due to the nature of biological material and partly due to the later development of high technology in biology than in the other disciplines, in spite of the ancient history of biotechnology.

Hazard. Risk assessment of biotechnological activities concerns governmental regulatory bodies throughout the world. All of the concerns are environmental and may be in such forms as a specific disease of humans, animals, or plants, an undesirable imbalance in the environment, or production of an undesirable product.

The data available to answer queries in the area of risk assessment are quite sparse. The most common category of useful data is the pathogenicity of strains. This is largely a strain-specific phenomenon, the degree of virulence varying from strain to strain, especially on serial propagation. Less commonly available are data describing strain persistence in various environments and toxicity of products and rarest of all perhaps are data to predict the effect of the introduction of strains into new environments.

2.2.5 *Taxonomic data*

The traditional and important method of constructing databases in service culture collections is with taxonomic orientation. The storage of the data and their public presentation considers the strains first as representatives of their taxa. Further, the data given in the description of each strain in the catalogue of collection holdings are sparse as they depend on the assumption that the reader knows, or is adept at finding, the usual attributes of the taxon. Many clinical microbiology laboratories only save the phenotypic data on antibiotic resistance patterns and the putative name of the isolates; the data used to decide the name are discarded. The records may indicate that the isolate is 'atypical' without any indication of the attributes that led to this description.

Given the traditional organisation of culture collection catalogues, the most common questions asked of service culture collections are likely to be on specific attributes of strains listed in the catalogues, or the companion question on the availability of strains with specific attributes. Indeed, experience shows that service collection personnel quickly develop great skill in searching the collection database by any and all means available to answer such questions from the community. In turn, the community learns that the collections are a valuable source of various kinds of information beyond that contained in the catalogues.

The taxonomic orientation is natural in managing diverse culture collections and determining the boundaries of interest for many specialty collections. Within a large collection, curator responsibilities are usually assigned along taxonomic lines.

Historically, many service collections were established for studying or supporting the study of taxonomy. This essential function still underlies a great deal of service collection activity.

Many of the laws and regulations concerning shipment, hazard, standards, laboratory safety, patents, use of biological material are stated all or in large part in taxonomic terms. The US Federal Register specifies strain and species in defining which strains are to be used as standards for antibiotic testing. The attributes of these strains are not listed. Similar regulatory documents exist for other parts of the world.

2.2.6   *Regulations*

The development of regulations for efficient and safe use of biotechnological processes to the common good is heavily dependent on adequate data of diverse types. Unfortunately, databases do not exist to allow detailed appraisal of the potential hazard on a case-by-case basis. The use of taxonomic levels to evaluate and regulate is contraindicated by the very nature of the process of establishing taxa. The only solution to this dilemma requires the gathering and dissemination of appropriate data.

2.3   **Information resources**

An informal infrastructure of information gatherers, managers, and disseminators exists to answer questions on the practice and regulation of biotechnology as it relates to microbial strains and cell lines. It forms a very useful resource in spite of its informality. Further, a number of initiatives are under way which aim to manage this support system for biotechnology in a more formal and complete fashion.

The infrastructure exists independently of the practice of biotechnology since the same needs for information transfer are basic to all of microbiology and cell biology. Collection holdings are raw material for these sciences and new information initiatives are stemming from the application of molecular biology and advances in biotechnology. These developments are merely refining the information pathways that evolved before modern genetic engineering focused the public eye on one of the oldest forms of manufacturing, the use of biological materials in processes.

The historical sequence of development of the infrastructure started with the culture collections, proceeded with catalogue production, followed by collections of strain data in computers and, most recently, the creation of national, regional, and finally international data services.

The ultimate source of data needed by the biotechnologist is the laboratory records of the collections. All other elements of infrastructure function to make these data accessible. The entry points to the pathways at all levels are the same as those considered previously: taxonomic or by detailed pattern of attributes. A combination is possible. The question 'Do you have a pseudomonad that degrades hexadecane?' will eliminate all yeasts that have the same ability.

2.3.1 *Culture collection catalogues*

Service collections publish catalogues describing their available strains to inform the public of their holdings and the salient properties of strains. A secondary effect is to reduce some of the labour overhead involved in answering questions from the public. A number of nations (such as Japan, China, and Brazil) have prepared combined catalogues, eliminating the need to obtain and consult multiple catalogues.

By the very nature of printed catalogues they function imperfectly since the information describing each strain is limited. Only one or at best a few attributes can be indexed so that detailed searching is impossible. Because of the positive accessioning policies of the service collections, catalogues are out of date upon publication. In general, the taxonomic entry route is served well, but the attribute route is necessarily left to follow-up questions to the collection itself. The most important deficiency is that only a very small proportion of the world's collections publish catalogues at all, since provision of a public service is not their prime function.

Where catalogues do exist, they form an important resource for the biotechnologist. The information they contain is carefully presented. If a taxon is known, much time can be saved by consulting a catalogue for availability. Often, valuable ancillary information on use, literature references, propagation conditions, or patents is included in the catalogue. Finally, the staff of the collection listed in the catalogue can be contacted for further information on their holdings or as entry into the rest of the informal information services of the collection community.

2.3.2 *Individual collections*

Culture collections of importance in biotechnology are not limited to the recognised service collections. In many cases, the biotechnologist must have access to a detailed collection of strains with the final selection of the particular strain for use in the process decided by personal comparative testing. Service collections may serve in this respect, but because of their usually broad nature cannot always maintain the specified holdings of the personal research or survey collection.

The mechanics of obtaining information from primary records of individual collections may not be simple. The curator must scan tables or individual strain records to match the pattern of the query. The process is faster for those collections which keep their records in a computer. While computer-aided searching is faster, considerable time still is required of collection personnel to search on a query by query basis. Time spent in searching the database to answer public queries may be considered a normal and reasonable function by the service collections or a burdensome chore by collections with other basic missions. Either way, such searches can represent a considerable overhead on staff time.

2.3.3   *Strain data compendia*

The availability of computers for both record-keeping and data analysis has resulted in compendia of strain data in locations other than the collections. Hospitals have contributed antibiotic resistance data to large-scale surveys of changes of resistance plasmid distributions in bacterial populations. Numerical taxonomists assemble large databases of strain data in the course of their activities, often containing data contributed by other research workers. Ecologists and public regulatory agencies conducting surveys of the environment, including habitats such as soils, waters, foodstuffs, and wild and cultivated plants, often amass considerable amounts of data which are installed in some computer resource for management and analysis. The result is that the data describing the strains reside in a different location from the strains.

Arising from all this activity is a body of curators of data in support of the curators of strains. The relevant information specialists may be associated with a collection, and where such data management exists, the biotechnologist's search is immensely facilitated. Since the taxonomic designation is managed in the computer in the same way as any attribute of the strain, the entry into the database can be taxonomic or by attribute pattern with equal facility. The problem of finding the strains of interest is largely reduced to the problem of finding the databases themselves.

2.3.4   *National, regional, and international data resources*

Scientists and technologists band together in organisations focused on their disciplines in order to exchange information. Many of these scientific and technical societies also become providers of services to their members and the public community. Some resemble guilds or unions in providing advocacy for improving conditions for their members. Recently, societies have come full circle in that they are providing, directly or through advocacy, informational resources in the form of publicly available databases.

Since most societies are national, the earliest of these database efforts were national as well. In some disciplines, national efforts were deemed so valuable that they became international in use. The Chemical Abstracts Service of the American Chemical Society is a well known example. Geopolitical regions have their counterparts in scientific and technical activities. The most notable of these regional efforts is within the European Economic Community with a ripple effect to other countries in Europe.

Either by combining regional resources or by direct international efforts, world-wide database resources are being established in many disciplines. Some stand alone and others are distributed in networks.

Informational resources describing culture collections and their holdings are distributed through all three levels as are the organisations concerned with sponsoring such resources.

National and regional organisations concerning culture collections. Microbiologists having interests in culture collections have banded together in national federations for culture collections. In some cases, cell biologists also are included. The boundaries for membership seem generally loose. The countries that have such national federations are listed and further information is provided in Chapter 8. The countries are: Australia, Brazil, Canada, China, Czechoslovakia, Japan, Korea, New Zealand, Turkey, United Kingdom, and United States of America. Japan, Brazil, and the UK are most actively pursuing national databases on collection holdings. Others are being discussed or planned. The Japanese database is designed to contain information from all types of microbial collections. The Brazilian and British systems are initially designed to concentrate on service collections. The systems limited to service collections could be expanded to include other collections as resources become available. More details on the data systems in the context of national and regional systems are given below.

Brazil
The second edition of the *Catalogo Nacional De Linhagens* was produced by the Fundação Tropical de Pesquisas e Tecnologia 'André Tosello' in Campinas, Brazil. The species names and designations of strains held by

23 Brazilian collections are listed in the catalogue. The catalogue includes bacteria, filamentous fungi, yeasts, protozoa, algae, animal cell lines, viruses, and miscellaneous unidentified microorganisms.

The database of the catalogue is being installed on the Brazilian national information system Embratel. Thus, it will be accessible through international telecommunication systems.

> *For more information contact:*
> Fundação Tropical de Pesquisas e Tecnologia 'André Tosello'
> Rua Latino Coelho, 1301
> 13.100 – Campinas – SP
> Brasil
> (Electronic Mail DIALCOM 42:CDT0094)

Canada

Curators of eight major culture collections holding fungi in Canada have contributed data to a common system maintained at the Atlantic Regional Laboratory of the National Research Council. Data on 19 500 strains are included in the following categories: name, accession number, substrate from which isolated, year of isolation, maintenance, literature references to metabolite production, pathogenicity, availability, and source. Collaborating collections include the Canadian Collection of Fungus Cultures (CCFC), Forintek Collection of Wood-Inhabiting Fungi (EFPL), University of Alberta Microfungus Collection and Herbarium (UAMH), and the National Research Council. Availability of data is currently restricted, but general on-line access is envisaged for late 1987.

> *For further information, contact:*
> Atlantic Regional Laboratory
> National Research Council
> 1411 Oxford Street
> Halifax
> Nova Scotia B3H 3Z1
> Canada

European Culture Collections' Organisation

The European Culture Collections' Organisation (ECCO) is an active group comprised of major service collections in countries that have microbiological societies affiliated with the Federation of the European

Microbiological Societies (FEMS). All the ECCO collections are also affiliated with the World Federation for Culture Collections. Thirty-three collections are members from Belgium, Bulgaria, Czechoslovakia, Finland, France, Federal Republic of Germany, German Democratic Republic, Hungary, Netherlands, Norway, Poland, Portugal, Spain, Sweden, Switzerland, Turkey, United Kingdom, and the United Soviet Socialist Republic.

ECCO was established in 1982 to collaborate and trade ideas on all aspects of culture collection work. Since service collections inherently are organised repositories of information as well as cultures, ECCO is a valuable resource for finding information of interest in biotechnology. Each of the collections produces a catalogue of holdings. Such catalogues often hold information beyond the listing and description of the strains held. Also, the curators are likely contacts for other, non-service collections in their countries.

As intercollection communication pathways grow and become formalised within and between ECCO countries, the prospects for an electronic communication network encompassing all these countries grows as well. Such a network is being actively discussed among its members at this time.

> *For further information contact:*
> European Culture Collections' Organisation
> Czechoslovak Collection of Microorganisms
> 662 43 Brno
> tr. Obrancu miru 10
> Czechoslovakia
> (Electronic Mail TELECOM GOLD 75:DBI0154)

Japan Federation for Culture Collections
Since 1951 the Japan Federation for Culture Collections (JFCC) has promoted cooperation among culture collections and individuals in Japan as well as internationally.

In 1953, the JFCC started collecting data on the holdings in Japan and completed the catalogue which listed about 22 000 strains from 144 research institutions. The strains in the list were reidentified by a project team organised by Professor Kin'ichiro Sakaguchi of the University of Tokyo, resulting in the publication of a series of JFCC catalogues in 1962, 1966, and 1968. Further publications are planned, with the next in the series for 1987.

For further information contact:
JFCC
NODAI Research Institute Culture Collection
Tokyo University of Agriculture
1-1-1 Sakuragaoka, Setagaya-Ku
Tokyo 156
Japan

Microbial Resource Centres (MIRCENs)
Information on Microbial Research Centres (MIRCENs) will be found in Chapter 8. These centres are found in both developed and developing countries. Each has its special focus of interest, for which it acts as a regional centre.

National and regional data resources

Institute for Physical and Chemical Research
In Japan, the Institute for Physical and Chemical Research (RIKEN) carries out international activities, such as being the host institution for the World Data Centre for Microorganisms and a node of the Hybridoma Data Bank. These activities are carried out by the Life Science Research Information Section (LSRIS) and Japan Collection of Microorganisms (JCM).

On a national level, LSRIS has developed the National Information System of Laboratory Organisms (NISLO), a directory of Japanese collections and their holdings. The NISLO covers microorganisms, animals, and plants. In the case of microorganisms, the LSRIS closely cooperates with the JCM.

The following information and services are currently available from LSRIS:

(1) the number of laboratory animals used in Japan and their scientific names;
(2) microorganisms maintained in the member collections of JFCC;
(3) algae maintained in culture collections in the world;
(4) identification of deciduous trees;
(5) fundamental references for cell lines widely used in Japan;
(6) bibliographical information for plant tissue and cell cultures.

For further information contact:
LSRIS
RIKEN
2-1 Hirosawa, Wako
Saitama 351-01
Japan
(Electronic Mail DIALCOM 42:CDT0007)

Microbial Culture Information Service (MiCIS)
One of the first (along with Japan) national data efforts is that of the Microbial Culture Information Service (MiCIS) in the UK. This initiative is oriented towards providing public access to primary observation data while retaining the taxonomic orientation for computer entry of the data. The initiative is a cooperative effort between the UK Department of Trade and Industry (DTI) and the UK Federation for Culture Collections. The following description is quoted from a joint statement of purpose between MiCIS and MINE (see below).

> This database has been developed by the Laboratory of the Government Chemist (LGC) on behalf of DTI following consultation with industry. MiCIS initially contains all data currently available on strains including catalogue information, hazards, morphology, enzymes, culture conditions, maintenance requirements, industrial properties, metabolites, sensitivities and tolerances. Further categories of information will be added as the system develops. MiCIS will contain data from all UK national culture collections and discussions are currently taking place to include data from private and other European collections.
>
> . . . Subscribers will be able to use MiCIS on-line or via a postal or telephone enquiry service and will pay an annual subscription plus a charge related to use. The system allows subscribers to search in confidence for the source of a named organism, the known properties of a named organism and for unknown organisms displaying particular properties. MiCIS News, a quarterly newsletter reporting MiCIS and culture collection activities, is supplied free to all subscribers.

Further information on MiCIS may be obtained from:
Microbial Culture Information Service (MiCIS)
Laboratory of the Government Chemist
Cornwall House
Waterloo Road
London SE1 8XY
UK
(Electronic Mail TELECOM GOLD 75:DBI0015)

Microbial Information Network Europe (MINE)
Within the European Economic Community (EEC) the regional activities
are primarily focused within the programmes of the Commission of the
European Community (CEC). They fund a number of initiatives, either
as sole efforts or collaboratively when the initiative has a scope beyond
the confines of the EEC. One such initiative within the EEC is the
Microbial Information Network Europe (MINE). MINE has the traditional taxonomic orientation of the service collections. The UK node has
provided the following description of MINE in a joint (with MiCIS)
statement of purpose.

> This EEC database is a computerised integrated catalogue of
> culture collection holdings in Europe and is prepared as a part of
> the EEC Biotechnology Action Programme. CMI [CAB Interna-
> tional Mycological Institute], in collaboration with CABI [CAB
> International] Systems Group, is to act as the UK node, in
> parallel with national nodes being developed in The Nether-
> lands, Germany, and Belgium; discussions with Portugal and
> France are currently under way.
> . . . it will not include full strain data but only a minimum data
> set.
> Once enquirers locate a culture, they will then be referred to
> the collection concerned, or to national strain data centres . . .
> for any detailed strain information if this is needed.
> Initially, enquirers will contact MINE nodes by mail, telex, or
> telephone. At a later stage, on-line services may also be avail-
> able on subscription.

Further information on MINE may be obtained from:
Direction Biology, Radiation Protection, Medical Research
SDME 02-41
(DG XII/F/2)
Commission of the European Community
200 rue de la Loi
B-1049 Bruxelles
Belgium
(Electronic Mail TELECOM GOLD 75:DBI0004)

Nordic Register of Microbiological Culture Collections
In 1984, the Nordic Council of Ministers initiated support for a Nordic
Register of Culture Collections encompassing Denmark, Finland, Ice-
land, Norway, and Sweden. The Register's scope is inclusive of all sizes
and functions of collections from small personal collections to large
service collections. The first three years of development is focused on
strains of importance to agriculture, forestry, and horticulture. The
building of the microcomputer-based database is an undertaking of the
Department of Microbiology of the University of Helsinki, Finland.
Development of software has been carried out in co-operation with the
Nordic Gene Bank for Agricultural and Horticultural Plants in Alnarp,
Sweden.

For more information, contact:
NORCC
Department of Microbiology
University of Helsinki
SF-00710 Helsinki
Finland
(Electronic Mail DIALCOM 42:CDT0069)

International resources
The primary focus for international culture collection information
resources is through the International Council of Scientific Unions
(ICSU) with headquarters in Paris, France. Various components of ICSU
have current or potential initiatives relating to providing information of
interest to biotechnologists. These include: Committee on Data for
Science and Technology (CODATA), World Federation for Culture
Collections (WFCC), International Union of Immunological Societies
(IUIS), and International Union of Microbiological Societies (IUMS).

Committee on Data for Science and Technology (CODATA)

The Committee on Data for Science and Technology (CODATA) was established in 1966 by the International Council of Scientific Unions (ICSU) to promote and encourage the production and international distribution of scientific and technological data. Its initial emphasis was in physics and chemistry, but its scope has been broadened to data from the geo- and biosciences. CODATA is 'especially concerned with data of interdisciplinary significance and with projects that promote international cooperation in the compilation and dissemination of scientific data'.

The main activities of CODATA are carried out by Task Groups established for specific projects. Of special interest to biotechnologists are the biologically oriented Task Groups on the Hybridoma Data Bank, Microbial Strain Data Network, and Coordination of Protein Sequence Data Banks. The first two are the policy boards of the activities while the last is a co-ordinating body among the existing sequence data banks.

> *For more information on CODATA contact:*
> CODATA Secretariat
> 51 Boulevard de Montmorency
> 75016 Paris
> France
> (Electronic Mail TELECOM GOLD 75:DBI0010)

Hybridoma Data Bank (HDB)

In 1984, the CODATA/IUIS Hybridoma Data Bank (HDB) started building an international database on hybridomas, other immunoreactive cell lines, and monoclonal antibodies. The HDB is designed to act as a locator service, help avoid duplication of effort, and provide a research tool on relationships among reactivity patterns. An international infrastructure (with data bank branches in the USA, France, and Japan) is in place, a growing database is being assembled, and queries are being answered. A pilot project for a publicly accessible on-line service is in progress.

The most common queries involve the antigen–antibody reactivity/non-reactivity patterns. The possible antigens to be named cover all of biology, biochemistry, and a large part of organic chemistry. The complexities compound rapidly when such problems as tumour epitopes are at issue. The host taxonomy, organ, tissue, cell structure, developmental stage, pathology, antigen, and epitope all have nomenclatural ambiguities to some degree. Various authority sources, selected by exercise of best judgement, used in building a controlled vocabulary/ glossary reduce the problems to manageable proportions. Communication paths, conventions, and frequent quality control exercises facilitate consistency of response.

The information on each cell line and product is quite comprehensive. In addition to reactivity attributes, considerable information is coded, such as fusion partners' histories, developer, availability, distributor(s), applications, assay procedures, antibody classes, immunisation techniques, literature and patent citations. Many of these categories have internal sub-categories.

Queries to the HDB may be through mail, telephone, or through the CODATA Network available on Dialcom which has nodes in 20 countries and links to the most common packet switching services as well as to the Telex and TWX systems.

On a pilot basis, a subset of the main database is available directly to the international public through the CODATA Network. The pilot subset emphasises antibody reagents readily available (commercially or otherwise) and their reactivities. Monitoring usage of this database yields a direct evaluation of the elements in the controlled vocabulary.

More information on the HDB is available from any of the three nodes:
Hybridoma Data Bank
12301 Parklawn Drive
Rockville, MD 20852
USA
(Electronic Mail DIALCOM 42:CDT0004)

LSRIS
RIKEN
2-1 Hirosawa, Wako
Saitama 351-01
Japan
(Electronic Mail DIÅLCOM 42:CDT0007)

CERDIC
Lab. d'Immunologie
Fac. de Médécine
Av. de Vallombruse
06034 Nice
France
(Electronic Mail TELECOM GOLD 75:DBI0014)

Microbial Strain Data Network (MSDN)

The CODATA/WFCC/IUMS Microbial Strain Data Network (MSDN) was started in 1985 to construct a world-wide network of holders of strain data serving as Nodes in an informational network. The MSDN will act as a locator service for strains of microbes or cultured cell lines having specific attributes.

From the UNEP-initiated round table in 1982 (at the International Congress for Microbiology) where the concept was introduced, to the present time, the Network has been designed; an infrastructure has been established, consisting of a policy-making Task Group, six operating Committees, and a Secretariat; and pilot on-line databases have been installed on a publicly accessible commercial computer host. Oversight is by three components of the International Council of Scientific Unions: CODATA, WFCC, and IUMS.

The sheer magnitude of the number of repositories of microbial strain data and the numbers of strains held within those repositories generate serious data acquisition and communication problems. Collecting the desired data in one or a few places for general availability is a practical impossibility. Rather, the MSDN is designed to operate as a locator service for repositories of strains with desired combinations of attributes by an indirect method. Thus, the data repositories become Network Informational Nodes.

The Central Directory of the MSDN contains a list of the data elements coded by all the various Nodes rather than the data themselves. This database is a controlled vocabulary of standardised nomenclature of mainly biochemical and morphological features used for strain characterisation throughout the world. The initial basis for the controlled vocabulary comes from the CODATA-sponsored publication by Rogosa, Krichevsky & Colwell (1986). The user scans the vocabulary to select features of interest. The features in the form of the controlled terms are entered as search criteria into a second database which yields the names of repositories assessing the strains for possession of the desired features. When a Node is located which codes information on the desired data elements (features), the person querying the MSDN contacts the Node(s) directly for existing strains fitting the detailed pattern of attributes. The contact may be accomplished through telecommunications (where the capabilities exist) or by mail or telephone. In an increasing number of cases, direct access to host computer databases is possible.

Queries to the MSDN may be through mail, telephone, or through the

CODATA Network available on Dialcom which has nodes in 20 countries and links to the most common packet switching services as well as to the Telex and TWX systems.

> *For further information on the MSDN, contact:*
> Institute of Biotechnology
> Cambridge University
> 307 Huntingdon Road
> Cambridge CB3 0JX
> UK
> (Electronic Mail TELECOM GOLD 75:DBI0001/DBI0005)

World Data Center on Microorganisms (WDC)

In 1984, the World Federation for Culture Collections (WFCC) officially reaffirmed the WDC as a component of the WFCC and thus accepted responsibility for the operation and management of the WDC. Because of the announcement of the retirement of the director of the WDC at Brisbane, Australia, a public search was conducted for potential hosts to ensure continuity of the WDC effort. After competitive evaluation of the proposals, the Executive Board of the WFCC agreed that the WDC be transferred to the Institute of Physical and Chemical Research (RIKEN), Saitama, Japan.

WDC is an information centre which supports culture collections and their users. The primary tasks are:

(1) publication of descriptions of culture collections;
(2) production of a species-oriented directory of culture collection holdings.

The tasks of WDC are not limited to the above; other tasks will be performed based on necessity and available resources.

Information sources for WDC are culture collections and national/local/international data centres as well. The WDC will, for example, make use of, and cooperate with, HDB and MSDN (as it becomes operational).

For information dissemination, the WDC will use a variety of communication media such as mail, telex, cable, facsimile, electric mail, publication and magnetic devices (floppy disks and magnetic tapes).

The WDC currently holds information on 327 culture collections distributed over 56 countries. The core data of the WDC are descriptions of culture collections and their holdings, allowing users to locate a culture collection and/or taxonomic category of microorganism. They can find strains of specific species by consulting the list of species preserved in culture collections. The WDC database currently includes bacteria, fungi, yeasts, algae, protozoa, lichens, animal cells, and viruses.

The WDC is able to answer queries by any communication media except on-line retrieval. The WDC will provide on-line information retrieval in the future, depending on available resources.

> *For further information on the WDC, contact:*
> WDC/RIKEN
> 2-1 Hirosawa, Wako
> Saitama 351-01
> Japan
> (Electronic Mail DIALCOM 42:CDT0007)

World Federation for Culture Collections (WFCC)

The World Federation for Culture Collections (WFCC) provides information to biotechnologists in a variety of ways within the overall mission of the WFCC as described in Chapter 8. Specific information initiatives described above are the World Data Center on Microorganisms and the Microbial Strain Data Network. Specialist committees established by the WFCC can provide information on a number of topics. The Patents Committee considers patent conventions. The Postal and Quarantine Committee is a good source of information on regulations governing shipment of cultures. The Publicity Committee publishes a newsletter on a periodic basis. The Committee on Endangered Collections identifies and takes action to rescue jeopardised collections. The Education Committee initiates training activities (e.g. books, videotapes, courses, individual instruction).

For more information on the activities and information resources of the WFCC contact the WDC (see above), the MSDN (see above), or any of the resource centres listed in Chapter 1.

2.4   Access to data resources

2.4.1   *Traditional*

Traditional methods for accessing data resources are still the most common and are quite satisfactory as long as the answers are not voluminous and are not needed quickly. Asking a culture collection curator about the availability of a single strain exemplifying a particular taxon or having a particular set of attributes will usually get a prompt reply. The query may be verbal, in person or on the telephone, or by mail. As the queries become more complex, the use of these methods become less satisfactory to both the seekers and providers of information.

The seeker of information becomes frustrated at the delay and incomplete nature of the answer that comes back. Often multiple cycles are required to refine the question to the point where the desired answer is given. While this refinement of communication is valuable in clarifying the true nature of the query (not always recognised from the initial inquiry) and common to all pathways, the frustration is amplified by the length of the cycle time.

The provider of information must devote increasing resources to answering queries. This takes professional expertise that could be used in the other work of the collection. Therefore, any mechanism which minimises the labour involved in answering queries increases the professional resources of the collection.

Both internal and external mechanisms are useful to alleviate some of the workload. Internal mechanisms include publications, such as the aforementioned catalogues, and computer management of the data for ease of searching and reporting. External mechanisms are primarily electronic forms of communication and are discussed in the next section.

2.4.2   *Electronic*

Except for voice communication via telephone, telecommunication has only recently become a part of collection life. Many of the larger collections have been using printed electronic messages (cables, Telex) for a while. However, the advent of computer-operated message transfer systems at reasonable cost (e.g. direct on-line access, electronic mail systems, and public packet switching services) have allowed economically practical electronic communication between questioner and answering resource.

Two parallel paths of development are taking place at this time in providing public access to data in collections. Some collections are following both paths simultaneously.

In the first instance, a collection may make its data accessible to the public by establishing access to a computer system maintained by the collection or its parent institution. Some examples are the Human Gene Probe Bank at the American Type Culture Collection, the National Collection of Yeast Cultures at Norwich, UK, Japan Collection of Microorganisms, Tokyo, and the CAB International Mycological Institute, Kew, UK.

The second approach is to install the data in a computer operated by others. The data and their installation may be accomplished under the control of the collection as is being done by the ATCC on the CODATA Network. Alternatively, all or part of the data may be installed on a computer operated as a national or regional facility such as those described above for Europe, Brazil, and Japan (MiCIS, MINE, Catalogo Nacional De Linhagens, and NISLO).

In all these cases of providing access through electronic computer services, the source of the data does not maintain the communication paths beyond the host computer. It is generally the responsibility of the seeker of information to find the most appropriate path. Unless the provider and seeker of information are at the same institution, where direct connection to the computer may be possible, the ordinary telephone system is likely to be the first resource connected to the seeker's terminal. Where short distances are involved, it may be reasonable to use only the voice-carrying telephone system. However, national and international data communications are using 'packet switching service' (PSS) for an ever-increasing share of data telecommunications. Such PSS transmission is cheaper and more reliable by far than direct telephone calls.

To find out more about telecommunications with the systems described in this chapter, get in touch with the appropriate system listed.

Help with establishing electronic communication paths may be obtained from the MSDN (see above).

3

Administration and safety

B. E. KIRSOP, C. P. KURTZMAN and K. ALLNER

3.1   Introduction

The different kinds of culture collections are described in Chapter 1, and from this it will be clear that the administrative procedures for the deposit and distribution of cultures will vary, depending on the role the collection plays. Those established primarily to fulfil a service function to the microbiological community will endeavour to recover some of the costs of maintaining the resource centre. Those acting more in an educational or research capacity will adopt a less commercial approach, since much of the benefit derived from their activities will be to the advantage of the parent institutions.

Many of the administrative activities required of culture collections are now computerised and lead to increased efficiency in the handling of requests and enquiries and to the streamlining of procedures for the day to day running of the collections.

3.2   Supply of cultures

Culture collections serve as repositories of living microbial cultures and, in some cases, of plant and animal cell lines. While this is in contrast to the preserved specimens found in herbaria and zoological collections, both types of repositories have the aim of preserving, cataloguing and making available for study the biological diversity found in the earth's flora and fauna. As such, these organisations serve as custodians, and frequently centres of taxonomic excellence, for specimens gleaned from the studies of a large number of researchers whose work represents a major investment of resources by society. Maintenance of these cultures and specimens for future reference spares us the expense of repetitive collecting and allows for the building of scientific knowledge on an ever-increasing base of information that is readily available for further examination and retesting.

3.2.1   *Location of strains*

To find out which strains are available from culture collections requires consulting the catalogues of particular collections (or the curator if no catalogue is issued) or an appropriate microbial database. The various pathways that can be followed to locate particular strains by name or attribute are described in Chapter 2. Strains for special applications are frequently noted in catalogues but recourse to the scientific literature may also be required since applications are not necessarily listed. On the basis of experience, collection staff are often able to predict which strains may be useful for various processes.

3.2.2   *Conditions of supply*

Cultures listed in catalogues are supplied on request and without restriction as to their use. By contrast, the supply of patent cultures is governed by various national and international regulations and laws and these are set out in Chapter 6. Patent cultures and those maintained for safe-deposit (Chapter 7) are administered in confidence, and the supply of such strains is restricted and only takes place under conditions agreed at the time of deposit.

Some collections maintain additional strains to those in the above categories that are held in reserve collections. These have usually been isolated by the staff at the collections for their own purposes and may be made available to others under special terms. Cultures such as these may be of interest for screening purposes, but as they are not listed in catalogues they can only be located by direct application to the culture collections.

Culture collections generally restrict distribution of strains to those making applications in an official capacity and who can be expected to be familiar with microbiological procedures (see Section 3.6.2). This is especially the case for those strains that are potentially pathogenic. The usual procedure to request a culture is by letter, telex, electronic mail or by telephone if verbal orders are accepted. Some guarantee of payment may be required if the collection charges for its strains, as most do. Methods accepted for ordering cultures from individual collections are listed in Table 1.2. Shipment of cultures, usually sent by mail, is governed by a host of regulations and these are discussed in Section 3.6.3.

3.2.3   *Pricing policy*

As with museums, the cost of maintenance of culture collections is not trivial. While one might suppose that collections could be self-supporting from the sale of cultures, such is not the case. One of the main reasons stems from the need of major collections to maintain a large diversity of microorganisms, many of which are requested only infrequently, but which nevertheless are considered important by the scientific community because they represent unexploited sources of germplasm which may achieve prominence in the future. For example, the recent exploitation of yeasts able to use pentoses, or of strains of brewing yeast able to ferment successfully in conical fermenters rather than traditional open vats, are instances of unpredicted uses for yeast strains previously maintained for academic interests only.

For large reference collections, sale of cultures seldom provides even half of the funds required for operation, and if actual costs had to be recovered through sales, the resulting prices would be prohibitive. The effect would be an ever-increasing exchange of strains between investigators that could lead to contamination, genetic change, or loss of strains. Although income from sale of cultures is often supplemented by charges made for other services (identification, preservation, patent deposit, safe deposit, contract research), the operational deficit of collections is almost always underwritten by grants from public funds. Such will probably always be the case, except for certain small commercial collections which specialise only in popular reference cultures and need not bear the maintenance costs of infrequently requested strains.

Culture collections may operate a differential pricing policy for the sale of cultures and other services, and the charges made to teaching, research or hospital organisations are usually subsidised. Some collections supply sets of strains for teaching at reduced rates. Collections almost always operate free exchange systems with other collections, providing that excessively large numbers are not requested, since it is felt important that strains of scientific significance are widely available. In addition, collections usually allow depositors free access to cultures of their own strains, within reasonable limits.

For up-to-date prices it is necessary to contact individual collections (see Chapter 1).

3.3   Deposit of cultures

Strains deposited in culture collections may be assigned to four categories: (1) cultures in the open collection that are available without restriction; (2) patent cultures, distribution of which is controlled by a variety of regulations; (3) safe deposit of strains maintained, usually in confidence, for particular clients who wish the convenience or security provided by the collection; and (4) strains held in reserve for the collections' own purposes. Additions to the general collection are made without charge and result from the contribution of novel isolates by various investigators as well as from solicitation of strains by the curators. Fees are generally charged for the other two categories. Some collections do not encourage deposit of numerous strains of the same species, whereas others may welcome such deposits if they differ significantly in some scientific or industrial manner.

Some collections do not accept particular species because of their special requirements for handling or containment. This being the case, depositors need first to inquire whether a collection will accept their particular strains.

3.3.1   *Accession procedures*

Collections generally require the completion of an accession form when strains are deposited. This may be a specially designed form drawn up by the collection and obtainable from them, or use may be made of the WFCC's SC4 form (Fig. 3.1). It is important that all information known about strains is transferred with the cultures, as the scientific value of undocumented strains is substantially less. The following list provides a guideline to data that should be supplied at the time of deposit.

Strain designation (number)
Taxonomic name (if known)
Origin – isolation (method, person, location, date)
History – previous strain designations, whether deposited in
    other permanent collections
Preservation requirements
Growth requirements
Applications
Pathogenicity/containment requirements

Once received, cultures are assigned an accession number and are checked for purity and authenticity. If these examinations prove satisfactory, the culture is formally accessioned and preserved using the

Fig. 3.1. The World Federation for Culture Collections' accession form (SCC-4).

WFCC Form SCC-4

NOMENCLATURE DATA

Genus	Species (include variety)		Type*
Saccharomyces	cerevisiae		Y

=100 Authority (Name and date)
Meyen ex Hansen (1883)

=110 World Directory Collection Number	Collection Acronym	Collection Accession Number
169	NCYC	738

HISTORY

=120 Received as (Genus)	Species, variety		=125 Date received
Saccharomyces	cerevisiae		Day 3 / 1 / 1972 Month / Year

=130 Received from (Name and address of person or collection)   As   =140 Strain Designation
P. Maule, Watney's Brewery, Mortlake, London, U.K.   -

=150 Who received it from (Name and address of person or collection)   As   =160 Strain designation
-

=170 Who received it from (Name and address of person or collection)   As   =180 Strain designation

=190 Also held in following permanent collections (Include number)
-
-

ORIGIN

=200 Isolated or derived from (If plant, animal or protist give genus and species name)
Isolated from continuous fermentation plant.

=210 Anatomical part (If applicable)

=220 Isolated from (Country, nearest town and distance and direction there from — If possible give latitude and longitude in degrees, minutes seconds and altitude)
U.K., London

=230 Isolated by (Name)   =240 Date
    Day / Month / Year

=250 Identified by   =260 Date
-    Day / Month / Year

MAINTENANCE

=270 Method of Preservation (Please tick whichever is applicable) and = 280 Temperature of Storage

		°C	
As Original Source Material			Agar Culture			°C
Lyophilised Culture	X	°C	Liquid Nitrogen	X	-139 °C
Other (Please specify)				

=290 Preservation suspending medium   Lyophilisation – glucose/horse serum
Cryopreservation – 5% glycerol

=300 Growth Medium (name and reference or give details on separate page)   =310 Temperature
YM broth/agar   DIFCO 0711   25 °C

=320 Growth conditions (e.g. aerobic, anaerobic, special gas mixture, light)   Microaerophilic

=322 Incubation time   =327 Subculture period
72 h   -

SPECIAL FEATURES AND USAGE (e.g. for citric acid production. Type strain etc.)

=330
Killer yeast, responsible for death of production strain.

REFERENCES (Journal, volume, pages, year)

=340   Journal, Institute of Brewing, 1973, 79 137. P. Maule & P. D. Thomas
=990

* Type – B = bacteria. A = algae. F = fungus. Y = yeast. P = protozoa. AV = animal virus. BV = bacterial virus.
IV = insect virus. L = lichen. TC = tissue cultures.
PV = plant virus.

If space is inadequate to reply to any question, please submit additional information on separate pages, identified by the appropriate reference number (e.g. =310)

63

methods considered best for the culture. Most service collections will preserve strains by more than one method. Figs. 3.2 and 3.3 summarise typical accession procedures adopted by service collections.

The accession number is the primary unit of strain identification. Most collections consecutively number strains in the order of accession. Because names may change as taxonomy progresses, the strain number serves as the sole constant in the record system. For this reason, it is

Fig. 3.2. Typical accession procedures used by service collections.

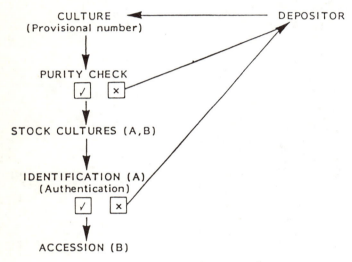

Fig. 3.3. Typical preservation procedures used by service collections. LN, liquid nitrogen; FD, freeze dried; QC, quality control.

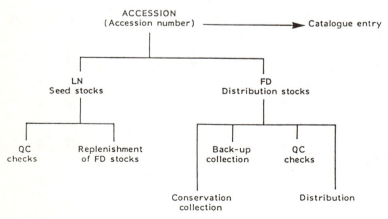

good practice never to reuse the number of a lost strain because of the potential for confusion. Strain designations comprised of letters or mixtures of numbers and letters may lend themselves to error more often than a consecutive numbering system.

The deposit of mixed cultures, that is those containing more than one strain of a species, or a mixture of species, presents problems for the collection because one member of the mixture frequently outgrows the other, making it difficult to maintain the special properties possessed by the mixture. Because of this, it is usually better to deposit each member separately and re-establish the mixture when needed.

3.4 Records

Accurate record-keeping is an essential and time-consuming part of the work of culture collections. At a minimum, records must be maintained on the following activities.

Accession (see Section 3.3.1)

Preservation –
   method
   calendar for maintenance
   survival/stability checks

Strain characteristics – morphology/physiology/genetical/
   industrial

References

Distribution –
   stock records
   customer records

3.5 Computer applications

3.5.1 *Catalogues*

The large quantity of data generated by culture collection activities lends itself particularly to computer management. For example, much of the accession data is published in lists or catalogues and most service collections now provide these with the aid of computers. The main advantages to be gained from this development are the ease with which catalogues can be updated and more frequently published, and the simplification of catalogue preparation, since the need for conventional and time-consuming proof-reading is eliminated.

3.5.2 *Strain data*

The very large amounts of data maintained by collections on individual strain properties is now frequently stored on computers, enabling easy updating and searching. Print-outs of all data on individual strains can readily be supplied when cultures are purchased, significantly enhancing the value of such strains to the scientific community. Collections are increasingly incorporating such data in publicly accessible databases (see Chapter 2), which can be searched, often on-line.

The literature references associated with strains are commonly stored in a format of the kind used for bibliographic databases. These, again, are easily searched by author, strain, date or subject keywords.

3.5.3   *Stock and customer records*

Collections supplying cultures need to maintain records on levels of stocks, particulars of customers, income and analyses of activities. This can best be achieved by an interactive system whereby the supply of a culture and the preparation of an invoice automatically depletes the appropriate stock levels and adds to the customer records. This enables regular records of current stocks to be obtained and a system set up for replenishment. Additionally, an analysis of distribution activities can be generated at regular intervals to record sales levels of individual strains, catalogues or other products, usage of the collection by different customer groups and overall income levels. Without this data a proper assessment of the value placed on the collection as a resource centre, and its development to meet future needs, cannot be made. It should be recognised that although such an analysis is important for predicting future trends and planning to meet them, it does not measure the overall scientific value of culture collections in providing microbiological information, training and research that cannot be quantified in this way.

3.5.4   *On-line computer developments*

Chapter 2 describes a number of developments for providing access to microbial data through computer networking systems. These are concerned primarily with accession data and strain properties, but culture collections can also use telecommunication networks to provide other services.

The National Collection of Yeast Cultures, for example, has developed a system that enables users to register with it and thus gain access to a number of on-line services that include ordering, catalogue viewing or searching, electronic mail, yeast identification (Chapter 5.2.5) and a 'news board'. These services may be used with equal ease by yeast workers anywhere in the world and only require access to the national or international computer networking systems. This may be obtained either using the minicomputers generally operating within universities, research organisations or companies (and frequently already linked to international bibliographic databases by computer networks), or by microcomputers linked to the telecommunications systems by modems (modulator/demodulators).

The provision of electronic mail facilities to collection users is particularly valuable, as it not only allows rapid and cheap communication between collections and users, but also between users themselves, thus creating an 'interest group' linked through the collection and providing the potential for increased collaboration.

3.6   Health and safety

Many culture collections maintain strains that are pathogenic or potentially pathogenic to humans, animals and plants. While the human pathogens are of immediate concern to staff members, all types of pathogens may have their impact on society. Although there have been no documented cases of accidental release of pathogens into the environment through mishandling in culture collections, this potential always exists and needs constant attention. Culture collection personnel must be aware of any pathogens that they may handle and the most expeditious means of containment.

3.6.1   *Laboratory practice*

Since the introduction of the Occupational Safety & Health Act 1970 in the USA and the Health & Safety at Work Act in 1974 in the UK, a spate of guidelines, regulations or recommendations have been produced pertaining to the handling of microorganisms in research and production establishments, hospitals and educational departments. There has been a conscious attempt to ensure that acceptable safety procedures are established and maintained at all places of work. In the USA in 1984, the Centers for Disease Control and National Institutes of Health have jointly prepared *Biosafety in Microbiological and Biomedical Laboratories*, which describes combinations of standard and special microbiological practices, safety equipment and facilities that constitute Biosafety Levels 1–4 recommended for working with a variety of infectious agents. The four classes and examples of organisms in each class are listed in Table 3.1. Table 3.2 outlines recommended laboratory handling procedures for infectious agents.

In the UK the Health & Safety Executive offers an advisory service to the community in all matters of safety and has inspectoral powers to enforce the law in instances where the Health & Safety at Work Act is breached. The latest guidance on the categorisation of pathogens is to be found in a report produced by the Advisory Committee on Dangerous Pathogens (ACDP) 1984, *Categorisation of Pathogens According to Hazard and Categories of Containment*. Bacteria, chlamydiae, rickettsiae, mycoplasmas, fungi, virus and parasites are clearly categorised according to the hazard they present to workers and the community, and four hazard groups are identified (1–4). Information is given on the degree of containment and protective clothing which should be applied during the handling of such organisms in the laboratory, including requirements for animal containment.

Table 3.1. *Classification of microorganisms according to biological hazard and their shipping requirements[a]*

Class I	Agents of no recognised hazard under ordinary conditions
Examples	*Saccharomyces cerevisiae, Trichoderma reesei, Lactobacillus casei*
Shipping	Culture-tube in fiberboard or other container. Permits as required
Class II	Agents of ordinary potential hazard
Examples	*Aspergillus fumigatus, Candida albicans, Cryptococcus neoformans, Staphyloccus aureus*
Shipping	Culture-tube wrapped in absorbent material, placed in metal screw-cap can, placed in fiberboard container. Permits as required
Class III	*Pathogens involving special hazard*
Examples	*Coccidioides immitis, Histoplasma capsulatum, Bacillus anthracis, Yersinia pestis*
Shipment	Culture-tube heat sealed in plastic, wrapped in absorbent material, placed in hermetically sealed can, placed in sturdy cardboard box. Permits as required. Etiologic agent warning label necessary
Class IV	*Pathogens of extreme hazard*
Examples	*Arthroderma simii, Pasteurella multocoida*, certain animal/plant viruses
Shipment	Culture-tube heat sealed in plastic, wrapped in absorbent material, placed in hermetically sealed can, placed in sturdy cardboard box. Required permits. Etiologic agent warning label necessary

[a] US Department of Health, Education and Welfare, 1972; US Department of Health and Human Services, Public Health Service, 1983.

Examples are:

Bordetella pertussis	Hazard Group 2
Candida albicans	Hazard Group 2
Cryptococcus neoformans	Hazard Group 2

Table 3.2. *Summary of recommended biosafety levels for infectious agents[a]*

Biosafety level	Practices and techniques	Safety equipment	Facilities
1	Standard microbiological practices	None: primary containment provided by adherence to standard laboratory practices during open-bench operations	Basic
2	Level 1 practices plus: laboratory coats; decontamination of all infectious wastes; limited access; protective gloves and biohazard warning signs as indicated	Partial containment equipment (i.e. Class I or Class II Biological Safety Cabinets) used to conduct mechanical and manipulative procedures that have high aerosol potential that may increase the risk of exposure to personnel	Basic
3	Level 2 practices plus: special laboratory clothing; controlled access	Partial containment equipment used for all manipulations of infectious material	Containment
4	Level 3 practices plus: entrance through change room where street clothing donned; shower on exit; all wastes decontaminated on exit from facility	Maximum containment equipment (i.e. Class III biological safety cabinet or partial containment equipment in combination with full-body, air-supplied, positive-pressure personnel suit) used for all procedures and activities	Maximum containment

[a] US Public Health Service, 1983.

Rickettsia prowazeki	Hazard Group 3
Naegleria spp. (especially *fowleri*)	Hazard Group 3
Lassa fever virus	Hazard Group 4

There are no yeast species listed higher than Hazard Group 2.

The American and British guidelines vary slightly but both have been accepted by the World Health Organisation. Other countries have developed similar systems or use those operating in the USA or the UK.

The control of genetic manipulation experiments in the UK is the responsibility of the Advisory Committee on Genetic Manipulation, using its own classification of experiments 1–4. Guidance is offered in a series of newsletters which are constantly updated.

The American control system is somewhat more complicated but all federal agencies that fund research related to biotechnology adhere to the policy that research in this field must conform to the requirements of the co-ordinated framework, such as the National Institute of Health Recombinant DNA Guidelines. Other countries engaged on recombinant DNA techniques have local guidelines or use the American or British guidelines.

3.6.2   *Supply of cultures*

Culture collections are aware of the responsibilities associated with the supply of cultures and require *bone fide* signatories before releasing hazardous pathogens. The National Collection of Type Cultures (NCTC) in the UK issues a leaflet to prospective recipients, explaining that all cultures supplied by them must be regarded as potentially pathogenic and be handled by or under the supervision of competent persons trained in microbiological techniques. This includes compliance with national or local codes of practice.

In the USA the American Type Culture Collection (ATCC) requires evidence that the recipient is trained in microbiology and has access to a properly equipped laboratory with the appropriate containment facilities. Requests for Class III pathogens must be accompanied by a signed statement assuming all risks and responsibility for subsequent use.

Customers ordering and receiving cultures from the German collection of microorganisms (Deutsche Sammlung von Mikroorganismen, DSM) assume all responsibility on receipt for the manner in which the cultures are handled and stored. The DSM does not accept liability for any injuries arising from the supply of cultures. Scientists within the German Federal Republic should consult the 'Vorläufige Empfehlungen für den Umgang mit pathogenen Mikroorganismen und für die Klassifikation von Mikroorganismen und Krankheitserregern nach den im Umgang mit ihnen auftretenden Gefahren' for details (see Further reading).

3.6.3   *Transportation of cultures*

The distribution of cultures, both domestically and overseas, is essential to further scientific endeavour world-wide. However, the shipment of microbial cultures subjects culture collections, as well as individual scientists, to a complex set of laws that have been designed to protect humans, plants and animals from disaster through accidental release or introduction of infectious agents. Most countries have laws concerning both the import and the export of microbial cultures, and frequently the states and provinces of these countries also regulate movement of pathogenic materials. Accordingly, those who ship or import cultures need to ensure compliance with these laws and regulations.

With regard to transport by air, the International Civil Aviation Organisation (ICAO) and International Air Transport Association (IATA) have reached agreement with curators of culture collections on procedures to be followed. Information is published by ICAO (1985) (see Further reading) providing details of the designation of dangerous goods, with packing instructions, which may be sent by air. A list of air operators holding general permission to carry such goods by air is also available. This information is continually updated and reviewed.

The postal authorities within the UK will also accept cultures for transportation either domestically or overseas. Conditions for the acceptance of Infectious Perishable Biological Substances (IPBS) must be sought from the Post Office Authorities prior to despatch, usually from a nominated Crown Post Office. Despatch of perishable biological substances in the overseas post is restricted to those countries whose postal administrations are prepared to admit such items. A list of countries accepting these items appears in the Post Office Guide, available from certain HMSO Offices (see Further reading). Senders are strongly recommended to ascertain from the addressee before despatch that the goods will be acceptable in the country of destination.

Infectious biological materials will be accepted by British Rail on a station-to-station basis (i.e. no collection service) either by a Recorded Parcels transit scheme or Red Star Premium Service. Terms and conditions using the United Nations Classification can be found in British Rail's publication (see Further reading).

In the UK the information on legislation governing the import and export of infectious biological substances can be obtained from either the Health and Safety Executive (HSE), Magdalen House, Stanley Precinct,

Bootle, Merseyside, or the Ministry of Agriculture, Fisheries and Food (MAFF), Hook Rise South, Tolworth, Surbiton, Surrey.

In the USA the importation of infectious biological substances capable of causing human disease is controlled by the Public Health Service Foreign Quarantine Regulations (42 CFR, Section 71.156). Importation permits, conditions of shipment and handling procedures are issued by the Centers for Disease Control. Similar restrictions and conditions for the importation of animal pathogens other than those affecting humans are subject to US Department of Agriculture regulations, Hyattsville, Maryland. The ATCC (1986) provides a summary of regulations affecting US scientists. The National Institutes of Health (NIH) have published a Laboratory Safety Monograph which provides additional information.

It is impossible to detail procedures for all countries here but reference to the following reading list should enable the reader to obtain the necessary information. Information on appropriate packing material and suppliers can be obtained through the service culture collections listed in Chapter 1. A number of culture collections, such as the ATCC, provide in their catalogues useful information on aspects of safety, or have advisory leaflets available for distribution.

There is no doubt that continued interest and concern will be expressed by both the scientific community and the general public about the safety aspects surrounding issues pertaining to microbiological agents. Safety professionals and scientists alike have an obligation to ensure that all microbiological agents are handled in a safe and responsible manner.

3.6.4   *Further reading (in chronological order)*

The Occupational Safety and Health Act of 1970. United States Occupational Safety and Health Administration. Public Law 91-596. 91st Congress S.2193. USA.

US Department of Health, Education and Welfare (1972). *Classification of Etiologic Agents on the Basis of Hazard.* Center for Disease Control, Atlanta, Georgia 30333, USA.

The Health and Safety at Work, etc. Act 1974, HMSO, London.

List of Dangerous Goods and Conditions of Acceptance by Freight Train and by Passenger train, BR. 22426 (1977 under revision), available from Claims Manager, British Rail Board, Marylebone Passenger Station, London NW1 6JR, UK.

Laboratory Safety Monograph: A Supplement to NIH Guidelines for Recombinant DNA Research (1978). National Institutes of Health, Bethesda, Maryland, USA.

Vorlaüfige Empfehlungen für den Umgang mit pathogenen
Mikroorganismen und für die Klassifikation von Mikroorganismen und
Krankheitserregern nach dem im Umgang mit ihnen auftretenden
Gefahren, published in *Bundesgesundheitsblatt* 23, 347–59 (1981).

US Department of Health and Human Services, Public Health Service
(1983). *Biosafety in Microbiological and Biomedical Laboratories*. National
Institutes of Health, Bethesda, Maryland, USA.

Categorisation of Pathogens according to Hazard and Categories of Containment.
(1984). Advisory Committee on Dangerous Pathogens, HMSO. ISBN
0.11.883761.3.

Biosafety in Microbiological and Biomedical Laboratories (1984). US Department of
Health and Human Services, Public Health Services, Centers for Disease
Control and National Institutes of Health, HHS publication No. (CDC)
84-8395.

The Air Navigation (Dangerous Goods) Regulations (1985). English language
edition of the International Civil Aviation Organisation. Technical
Instructions for the Safe Transport of Dangerous Goods by Air (DOC
9284-AN/905).

American Type Culture Collection (1986). *Packaging and Shipping of Biological
Materials at ATCC*, Rockville, Maryland USA.

LAV/HTLV III – the causative agent of AIDS and related conditions. Revised
guidelines (1986). Advisory Committee on Dangerous Pathogens, DHSS.
Health Publications Unit No. 2 Site, Manchester Road, Heywood, Lancs.
OL10 2PZ, UK.

Undated references

Advisory Committee on Genetic Manipulation. Guidelines and newsletters.
ACGM Secretariat, Baynards House, 1 Chepstow Place, London W2 4TF,
UK.

Conditions of Supply of NCTC Cultures: Hazardous Pathogens. Public Health
Laboratory Service, Central Public Health Laboratory, 61 Colindale
Avenue, London NW9 5HT, UK.

Importation Permits available from Centers for Disease Control, 1600 Clifton
Road, NE, Atlanta, Georgia, USA.

Public Health Service Foreign Quarantine Regulations, 42 CFR. Section
71-156.

The Post Office Guide, available from certain HMSO Offices, including 49
High Holborn, London WC1V 6HB, UK.

4

Culture and preservation

B. E. KIRSOP

4.1   General cultivation

Yeasts are facultative microorganisms, able to grow under both aerobic and anaerobic conditions when supplied with a source of metabolisable carbohydrate, nitrogen (organic or inorganic), minerals including trace elements, and vitamins. Almost all species can be grown on simple synthetic media, more complex synthetic media or media prepared from malt extract or brewer's wort. A list of suitable media is given in Section 4.2.2. These media may be prepared as a liquid or in a solid form by the addition of agar and most yeast species will grow well with either.

The pH range is not critical and, although the optimum range is between 4.5 and 6.5, pH values between 3.0 and 8.0 can be tolerated.

Temperature tolerances of from −2 °C to 49 °C have been recorded, the lower temperature commonly restricted to yeasts isolated from cold environments, but not always so. Yeasts isolated from man and animals are generally able to grow at higher temperatures, but again species such as *Kluyveromyces marxianus* (frequently isolated from dairy products) can grow at temperatures as high as 46 °C. Yeasts are, therefore, tolerant of a wide range of temperatures and information on strain variation is available from some culture collections. The ability of strains to grow above 37 °C is often used for identification purposes and most collections of yeasts will be able to supply this information to enquirers. The optimum growth temperature for many yeast species will be between 20 and 25 °C.

It follows from this that most yeasts are undemanding nutritionally and with regard to pH, temperature and oxygen requirements. There are a few exceptions, however. *Malasezzia furfur* (*Pityrosporum ovale*), for example, has an absolute requirement for myristic or palmitic acid, and *Cyniclomyces guttulatus*, isolated from the intestinal tract of rabbits, requires protein hydrolysate such as proteose peptone or trypticase for growth in addition to a high level of atmospheric carbon dioxide. Many yeasts have an absolute requirement for one or more vitamins, biotin and thiamine being commonly needed.

Osmophilic or halophilic yeasts are capable of growing in high concentrations of sugars or sodium chloride and may thrive best under these conditions. Most, however, will also grow well on general purpose growth media.

4.1.1 Scale-up

In order to avoid contamination by unwanted microorganisms, it is clearly important to ensure the sterility of all media and equipment and to use aseptic techniques in all manipulations. An exception is in the use of a very heavy inoculum, which will compete successfully with small numbers of contaminants.

A large-scale inoculum for industrial purposes must initially be obtained by growing a sufficient quantity of cells from a pure stock culture, usually a freeze-dried culture or an agar slope culture. This may be done using any convenient equipment, but stainless-steel propagation vessels are generally used. The culture is grown in successively larger volumes of substrate, under optimum conditions for the strain, until a sufficient amount of inoculum has been obtained. It is general practice to restrict each consecutive increase in volume to $10\times$ or less. If a small number of cells is inoculated into a large volume of substrate, growth is unlikely to be successful. It has been suggested that this is due to the inadequate level of carbon dioxide available as an essential nutrient for growth.

In order to grow a culture from a freeze-dried ampoule (containing about 10^6 viable cells), the following procedure could be used.

> Open ampoule according to instructions.
>
> With a Pasteur pipette, transfer enough growth medium from a container holding 10 ml to wash out the dried material from the ampoule.
>
> Return the yeast suspension to the rest of the medium.
>
> Repeat the washing procedure if necessary.
>
> Incubate the 10-ml broth culture for 48 h at 25 °C; faster growth is generally obtained by shaking the culture to keep the yeast in suspension and increase the level of oxygenation.

Transfer the whole of the grown culture to a larger vessel containing 100 ml growth medium and incubate as before.

Transfer the 100-ml culture to a 1-litre vessel.

It may be found that somewhat greater scale-up can be obtained after the first two transfers, but it is particularly important to keep the first two steps at the 10× level.

The level of inoculum used in fermenters will depend on the time required for completion of the fermentation, the composition, oxygenation and temperature of the substrate and the physiological condition of the inoculum. In a typical brewery fermentation an inoculum level of 2.5 g wet weight of yeast per litre will provide an acceptable fermentation rate in an air-saturated wort.

4.1.2   *Isolation*

Successful isolation of yeasts depends on a great number of factors, such as species to be isolated, composition of the substrate, age and condition of the cells, whether all or particular species are to be isolated and the level of contamination by other microorganisms. Media may be selective for yeasts or differential for particular yeast species. Selective media may contain antibacterial or antifungal agents or both. For further information, the reader is referred to Beech & Davenport (1971), Beech *et al.* (1980), and van der Walt & Yarrow (1984*a*).

4.2   Preservation methods

4.2.1   *Introduction*

During the development of industrial microbiological processes early consideration should be given to the culture preservation system to be used for the maintenance of production strains. It is frequently the case that the importance of this aspect of the process is only recognised when the first production run fails and the question then arises whether this could be due to the failure of the inoculum.

Service culture collections are well aware of the need to establish reliable preservation protocols and have found that these may vary with genus, species or indeed strain (Kirsop, 1974). In biotechnology it is of crucial importance that optimum procedures be determined for important production strains so that high viability and uniform cultural performances are ensured. Fermentation management procedures are generally designed following trial runs using a standard inoculum; subsequent alterations in inoculum performance due to a gradual reduction in viability on storage or to strain drift following inappropriate preservation protocols can lead to serious production problems. In most cases the situation is retrievable by emergency rescue operations on the substandard inoculum; occasionally these tactics do not succeed and the culture is lost. In either case there is considerable disruption to production, and severe financial losses may ensue.

It is clear, therefore, that early attention should be paid to the selection of reliable storage methods; in the case of a strain that represents a high level of past investment and on which future production depends, it may be in the best interests of the company to take the precaution of depositing the strain with one of the service collections offering safe-deposit facilities (see Chapter 1, Table 1.1). If this is done, the culture will be maintained expertly, often by more than one method, on a strictly confidential basis; cultures will always be available to the depositor, or to persons designated by him, on demand. Although use of this service provides sensible back-up for valuable strains, it should not be used as a reason for failing to establish good in-house preservation procedures, since recourse to a culture collection for a replacement strain will inevitably cause some interruption to production.

The establishment of good culture preservation practices becomes of even greater importance when genetically manipulated yeasts are used, since these may be less stable than wild-type strains. This special problem is discussed separately in Section 4.3.

Because of the many factors contributing to strain survival, it is sometimes difficult to select the best method of preservation for particular circumstances. Factors to be considered can be divided into microbiological factors, related to the intrinsic characteristics of the strain to be preserved, or extrinsic factors such as equipment, manpower and expertise, fiscal constraints or climatic factors. These are discussed further in Section 4.4 and Table 4.3 is provided to help assess the relative importance of different criteria when selecting a preservation method.

Methods suitable for yeast preservation fall into five major categories:

Subculturing – serial transfer

Drying – desiccation

L-drying

Freeze-drying – lyophilisation

Freezing – cryopreservation

4.2.2 *Subculturing*

Introduction. Subculturing, or serial transfer, is an essential technique familiar to all practising microbiologists. It consists of the transfer of a sample of cells from spent to fresh medium so that the cells may continue to proliferate. The procedure may be continued indefinitely and be tailored to suit the requirements of the laboratory. When used as a system for the storage of yeasts, steps are taken to extend the period between transfers, since cell death of the total population will normally be complete within a few weeks. Cell viability can be extended for longer periods by reducing the metabolic rate. This can be brought about either by lowering the incubation temperature or by limiting access to oxygen.

There are significant advantages and disadvantages to the method, which under certain circumstances would have an overriding effect on the selection of subculturing as an appropriate preservation system. These are discussed further in the Section 'Advantages and disadvantages' below.

Media. Any general yeast growth media may be used (see below). Either broth or agar is suitable, although many ascomycetous yeasts more commonly form ascospores on agar, and if strain stability is important broth media may be preferred. However, inadvertant contamination of cultures is more readily detected on solid media.

Non-fermentative yeast species do not survive so well in broth cultures as on agar slopes. Species belonging to the genera *Cryptococcus, Rhodotorula, Sporobolomyces, Bullera, Lipomyces* and, occasionally, *Candida* should be maintained on both kinds of media initially to check survival levels. Strains belonging to species of *Brettanomyces* and *Dekkera* produce organic acids abundantly and this will usually reduce the shelf-life of stored cultures. More frequent subculturing of these strains or the addition of calcium carbonate to the medium will improve survival levels.

Strains belonging to *Kloeckera* or *Hanseniaspora* may also require more frequent subculture than normal, and the incorporation of added vitamins to the medium may be necessary. *Schizosaccharomyces* strains may similarly need more frequent subculturing.

Some yeast characteristics may only be maintained under selective growth conditions. For example, strains of *Zygosaccharomyces bailii* resistant to high levels of benzoate may lose this characteristic if subcultured on media without benzoate. Similarly, strains isolated from sea-water lose their halotolerance when grown on media with normal sodium chloride levels. Strains may lose plasmids unless appropriate selection pressures are maintained (see Section 4.3). These requirements should be taken into account when selecting a medium suitable for subculturing.

Media appropriate for most yeasts, or suitable for adjustment for strains with special requirements, are:

YM broth   (Difco 0711-01)

YM agar   (Difco 0711-02)

PDA   (Difco 0013-01-4)   Potato dextrose agar

Wort or wort agar

Sabouroud/dextrose agar

Potato dextrose agar is particularly suitable for basidiomycetous yeasts. Shelf-life of subcultured strains may be extended by the use of agar slants that have been covered with a layer of sterile mineral oil.

Subculturing may be carried out using a variety of containers. Test tubes (cotton wool plugged or loosely stoppered), screw-capped bottles (Bijoux, McCartney, Universal) or vials are appropriate. They must be well washed and sterilised before filling with medium.

Inoculation, incubation and storage. Fresh medium is inoculated from a grown culture. A double tube method should always be used, whereby two tubes are inoculated: one is used only once for the next subculture, and the other is used as a working stock culture.

Inoculated cultures are incubated at temperatures optimum for growth; for most yeasts this is between 20 and 30 °C, although growth tolerances range between −2 and 50 °C (see Section 4.1).

Incubation is continued until the stationary phase of growth is reached. For most yeasts this is about 72 h after the start of incubation; it has been found that storage of younger cells can lead to reduced survival levels.

Stationary phase cultures are transferred to appropriate storage containers, growth characteristics first having been recorded for quality control purposes. The temperature of storage must be low enough to reduce metabolic activity, and thus increase shelf-life, while not being so low as to allow ice-crystal formation. A temperature between 2 and 4 °C is frequently used, although some laboratories prefer higher temperatures of 4–10 °C.

Survival and stability. The length of time that can elapse between transfers depends on the medium, age of culture at storage and temperature of storage; it is, moreover, strain specific. It follows that the best time to subculture can only be estimated accurately by experimentation with individual strains. This is not always practicable, particularly where large numbers of cultures are maintained, and a protocol of wide application is generally adopted.

Transfer at 6-monthly intervals is a time-scale found to be appropriate for a wide range of yeasts. Genera for which more frequent transfer may be required include *Cryptococcus*, *Rhodotorula*, *Lipomyces* and *Brettanomyces* (see Section 4.2.1). In the case of the genus *Bullera*, for example, many cultures are reported to be dead after 2 months' storage; shelf-life may be extended if potato dextrose agar is used (T. Nakase, personal communication). It is reported that many cultures on agar slants with an oil overlay may be stored for as long as 2 years before transfer is required.

The viability of subcultured strains is generally assessed by whether the 'culture' continues to grow following transfer. This is a deceptive criterion, since 'growth' may occur even though 90% of the original population is no longer viable (von Rehberg, 1978). Most yeast cultures survive the subculturing procedures described above, but it has to be remembered that the inoculum used for each successive transfer may not represent the whole of the original culture.

It is recognised that the 'industrial activity' of strains is seldom at an optimum level following storage, and this factor must be recognised when planning culturing procedures for production strains (see Section 4.5).

If the stability of strain characteristics is of major importance. Subculturing is not the best method to use. There are a large number of references (Kirsop, 1974; Butterfield & Jong, 1975) describing strain drift in different cell types. In a study of 300 brewing yeasts, 10% showed different flocculation characters following 10 years' preservation by subculturing. Other studies show that morphological, physiological, industrial and genetic characters may be unstable over prolonged periods of maintenance by subculturing.

Advantages and disadvantages. Subculturing is a widely applicable method for the preservation of yeasts. It has short-term advantages in that it is technically simple, cheap in terms of equipment (although expensive in terms of labour) and versatile. It can be adapted for the preservation of fastidious species, or combined with stringent selection pressures for the maintenance of constructed strains (Section 4.3) or used to ensure high product levels by addition of appropriate precursors to the medium. It is suitable for distribution purposes if solid media are used.

However, the method is monotonous to carry out and mislabelling and misinoculation can occur. Steps must be taken to minimise these errors by, for example, randomising labelled bottles for inoculation (so that the operator must search deliberately for matched pairs of bottles) and adopting good quality control procedures. The most serious disadvantage of the method is, however, the low level of viability and high level of strain degeneration found to occur following prolonged subculturing.

The method has serious limitations for culture collections, where the long-term stability of strains is of particular importance. In industry, attempts should be made to establish the stability of production strains and if instability is suspected, alternative long-term procedures should be adopted.

4.2.3   *Desiccation*

Introduction. Preservation of microorganisms by drying, or desiccation, has been used widely but there is little documentation about the levels of viability and strain stability attained. Methods used, all of which are technically simple, include drying on sand, soil, kieselguhr, silica gel, paper or gelatin discs, plugs of starch or peptone. Because of the lack of quantitative data, it is necessary to carry out preliminary tests before applying them to strains not previously preserved by this method.

Yeasts have been dried using (a) purified anhydrous silica gel as a desiccant or (b) squares of Whatman No. 4 filter paper dried in a desiccator.

Preparation of desiccants. (a) Silica gel (mesh 6-22) is sterilised by drying in an oven (180 °C for 90 minutes) in screw-capped bottles that have been filled to a depth of 1 cm. The gels are stored in a dry, warm atmosphere before use.
(b) The filter paper is cut into 1-cm squares, which are placed, 4–5 to a packet, in 6-cm squares of aluminium foil. The foil is folded once and the packets are sterilised by autoclaving.

Inoculation and storage. (a) Gels are inoculated with several drops of a heavy suspension of a 72-h agar slant culture in autoclaved, 5% milk. Before inoculation the sterile gels and milk are cooled overnight in a refrigerator and placed in trays of ice throughout the inoculation procedure. This is to disperse any heat generated when the gels are moistened that may damage the cells.

The bottles of inoculated gels are screwed tightly and stored in airtight containers containing indicator silica gel at room temperature or in a refrigerator. Cultures are revived by shaking a few gels into an appropriate medium and incubating.

An alternative inoculation method, in which cells are taken from a well-grown slope culture and placed with a spatula on the inner wall of gel-filled screw-capped vials, has been used by Banno, Mikata & Yamauchi (1981).

(b) Filter paper squares are inoculated by immersing into drops of a yeast : milk suspension prepared as in (a). The squares are dried in a desiccator in once-folded foil packets, after which time the packets are closed and stored in an airtight container at 4 °C. Cultures are revived by removing a filter square and placing on an agar plate or in broth medium and incubating.

Survival and stability. Quantitative data are not available for survival levels of yeasts stored by either of the desiccation methods described above. It has been reported (Bassel *et al.*, 1977; Kirsop, 1984), however, that a number of species fail to survive following 2 years' storage on silica gel using the milk suspension method and that response to this preservation system is strain specific. It follows that survival of individual strains should be checked before depending on the method for long-term preservation. By contrast, Banno *et al.* (1981) report that using their method, 90.5% of 200 yeast strains survived 10 years' storage, although strain stability was not recorded.

Survival of yeasts on filter paper is more successful, particularly for genetically marked strains of *Saccharomyces* (Bassel *et al.*, 1977), although it may be unsuitable for other genera.

Stability of the properties of gel-stored cultures is again strain dependent, the brewing characteristics of some strains of *Saccharomyces cerevisiae* being seriously affected, whereas others appear unchanged (Kirsop, 1978). Genetic markers of cultures stored on filter paper appear to be stable, but there is little data on the stability of other characteristics.

Advantages and disadvantages. As with subculturing, desiccation methods are simple to perform and require no expensive apparatus. They are particularly suitable for laboratories with limited resources or situated in regions with a high ambient temperature. However, the sensitivity of a number of species to preservation by desiccation limits the usefulness of the technique. Additionally, surviving strains may exhibit altered physiological, genetic or industrial properties. The method should only be adopted after it has been established that the strains to be preserved are not adversely affected.

4.2.4 *L-Drying*

L-drying is a method used for the preservation of microorganisms in which water is removed directly from the liquid phase, rather than by sublimation from ice as in freeze-drying. The ampoules are attached vertically to a manifold and immersed in a water bath at 20 °C. The vacuum level is adjusted to allow evaporation without frothing or freezing. Primary drying may be completed within half an hour, after which time the ampoules are constricted and secondary drying and sealing under vacuum are carried out as for freeze-drying.

The method was originally described by Annear (1958) and has since been adapted by Mikata, Yamauchi & Banno (1983) for the preservation of yeasts, using phosphate buffer containing 5% monosodium glutamate, 5% lactose and 6% polyvinylpyrrolidone at pH 7.0 as the suspending medium. Of the 1710 strains from 385 species and 46 genera preserved by this method, 91% were estimated by an accelerated storage test to be likely to survive for 50 years at a storage temperature of 5 °C. Yeasts that were filamentous, osmotolerant or psychrophilic appeared to be more sensitive to L-drying.

4.2.5 *Freeze-drying (lyophilisation)*

Introduction. Freeze-drying is used for the preservation of a wide range of microorganisms with varying degrees of success. The process resembles desiccation in that water is removed from cell suspensions by drying, but differs from the methods described in Section 4.2.3 in that the water is removed by sublimation from the frozen material rather than by evaporation from the liquid suspension.

The freezing part of the procedure may be carried out separately from the drying, or the two activities may take place as part of a continuing process. A number of commercial machines are available for the purpose and may be of the shelf or centrifugal type. In the former system, freezing takes place as a result of lowering the temperature of the shelves; in the latter, freezing occurs as a result of rapid evaporation under vacuum.

The centrifugal procedure is generally considered to be more suitable for long-term storage, since in this method the ampoules are hermetically sealed, conferring a longer shelf-life, whereas the serum bottles commonly used in the shelf system are sealed with rubber seals which are not entirely impervious to air and water. Moreover, the ampoules used in the centrifugal system may be cotton wool plugged for added protection, a procedure not possible with serum bottles. Some shelf systems use glass ampoules that may be plugged and heat sealed after removal from the chamber.

It is possible to combine features of both systems by using a double-tube method (Alexander *et al.*, 1980). The inner tube is frozen in a refrigerator and then evacuated to allow drying to continue overnight. The inner tube is then placed in an outer tube (containing a little silica gel) which is evacuated and sealed as in the centrifugal method. This system involves more manipulations, but is more flexible.

Inoculation, storage and recovery (centrifugal method). A detailed description of commonly used procedures is given in *Maintenance of Microorganisms* (see Kirsop, 1984). Ampoules are prepared by washing, drying, labelling, plugging very loosely with cotton wool (or covering with sterile felt caps) and sterilising by autoclaving. The inoculum is prepared by mixing equal proportions of a culture of post-logarithmic cells and suspending medium. There are a number of protective substances that have been shown to be suitable for the purpose (glucose–horse serum, inositol–serum, inositol broth, skimmed milk, sucrose). Observations described in Section 4.2.5 are based on studies using a 7.5% glucose–horse serum suspending medium. The filled ampoules are placed in the primary drying chamber, which is then evacuated, allowing rapid freezing and preliminary drying to take place. After a period of 2–3 h, the vacuum is released, and the ampoules are removed, constricted and placed on a manifold to allow secondary drying to take place (Fig. 4.1). The purpose of constricting the ampoules is to allow them to be easily sealed after drying is completed and while still under vacuum. Ampoules stored under an inert atmosphere have an extended shelf-life.

The residual moisture content of freeze-dried material affects survival levels (Fry, 1966) and has been shown for many cell types to be optimum at about 1%, depending on the suspending medium used. There are no data available on the effect of residual moisture content on the survival of different yeast species, but in practice an overnight secondary drying programme appears suitable for most yeasts.

Constriction can be carried out either manually, using a gas–air or gas–oxygen mixture, or with a semi-automatic constriction machine.

Final sealing can be carried out with a hand torch, a domestic blow torch or a crossfire burner as used for constriction.

Ampoules are checked for vacuum leaks using a high-vacuum spark tester and if satisfactory are stored in the dark. It is preferable to store at a

Fig. 4.1. Secondary stage of freeze-drying.

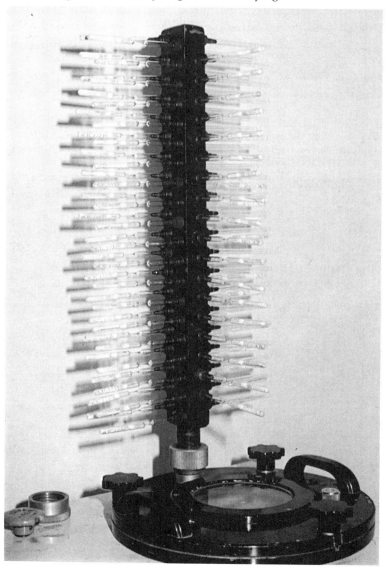

reduced temperature, particularly in tropical regions. Figure 4.2 shows different stages in ampoule manipulation during the freeze-drying process.

Cultures are recovered by resuspending the dried material in an appropriate growth medium and incubating at an appropriate temperature (Section 4.1). It should be recognised that cells that have been freeze-dried require a longer incubation period than normal to show growth, and when schedules are being established for the preparation of industrial inocula, account should be taken of this.

Survival and stability. Although freeze-drying has been used for the preservation of yeasts for some time, it has long been recognised that

Fig. 4.2. Ampoules at different stages in freeze-drying. From top to bottom: washed ampoule; plugged and labelled ampoule; ampoule constricted following primary drying; sealed ampoule.

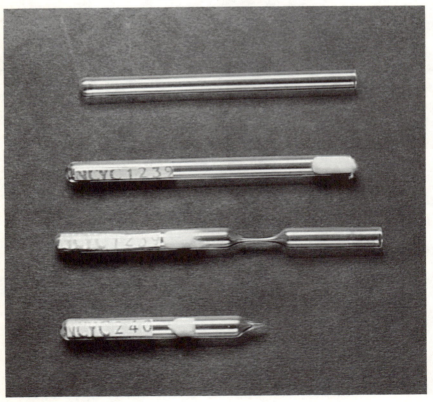

survival levels are low (Kirsop, 1974). Table 4.1 shows average percentage viability figures for different yeast genera. The differences shown between genera are not significant, since within any genus, survival levels vary and resistance to freeze-drying stresses is strain dependent (Fig. 4.3).

Attempts have been made to increase survival levels and it has been shown that post-logarithmic cells are more resistant than younger cells. Additionally, for most yeasts, growth without aeration leads to improved survival. Experiments to relate this to total lipid composition were unsuccessful (Kirsop & Henry, 1984), and factors other than the degree of saturation or unsaturation of fatty acids and sterols are clearly involved.

Growth of sensitive cells on a low-nutrient medium may improve survival levels, but the use of different suspending media and alteration

Table 4.1. *Survival levels of yeast genera following freeze-drying*

Genus	Mean % survival	Standard deviation	S.E.M.	No. of strains
Brettanomyces	2.21	2.53	.89	8
Bullera	3.7	3.05	1.76	3
Candida	12.95	19.1	2.45	61
Citeromyces	11.28	12.76	9.03	2
Cryptococcus	21.48	27.49	8.69	10
Debaryomyces	15.04	18.2	5.5	12
Dekkera	.35	.13	.10	2
Hanseniaspora	7.5	7.64	5.4	2
Hansenula	5.1	6.99	1.2	34
Kloeckera	25.44	24.74	8.75	8
Kluyveromyces	9.12	19.29	3.78	26
Lipomyces	5.29	5.46	3.15	3
Metschnikowia	11.19	8.64	2.88	9
Nadsonia	4.08	4.84	3.42	2
Pichia	9.10	15.54	2.63	35
Rhodosporidium	4.40	2.47	1.75	2
Rhodotorula	11.75	22.05	4.50	24
Saccharomyces	5.16	10.80	.45	580
Saccharomycodes	.01	.006	.003	5
Saccharomycopsis	11.36	10.4	3.83	7
Schizosaccharomyces	7.04	10.55	3.99	7
Sporobolomyces	1.4	1.37	.54	6
Trichosporon	8.27	11.56	4.72	6
Trigonopsis	.76	.91	.65	2

From Kirsop & Henry (1984).

of rehydration procedures have not been shown to affect survival significantly. Nevertheless, experiments have usually been carried out on a limited number of strains, and improved survival levels may be obtained with individual strains by judicious adjustment of pre-growth, freeze-drying and rehydration parameters. Although it has been shown (Alexander & Simione, 1980) that incubation of rehydrated cells at 0 °C increases viability and there is some evidence that dehydrated cells recover better if suspended in dilute rather than full-strength medium due to a slower and more orderly return to normal cellular activity, improvements obtained by these practices are small compared with the initial losses sustained during the freeze-drying process.

Long-term survival of freeze-dried yeasts is very good, extending to over 50 years in some cases. Nevertheless, it has been found that individual strains may abruptly lose viability, and it is prudent to monitor survival levels routinely, using either plate counts (Miles & Misra, 1938) or methylene blue staining (EBC Analytica Microbiologia, 1977) techniques.

The stability of freeze-dried yeasts has been exhaustively studied using cultures maintained at the National Collection of Yeast Cultures (Kirsop, 1974). It was found that variation in morphological, physiological and industrial properties was very much less than had been recorded

Fig. 4.3. Percentage survival of 117 strains of *Saccharomyces cerevisiae* after freeze-drying.

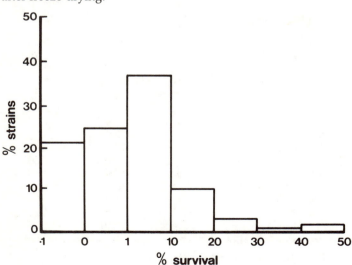

for equivalent subcultured strains. Changes in general fermentation characteristics such as flocculation, speed and degree of fermentation, and oxygen requirement to reach normal completion of fermentation were not detected in selected production strains. However, some genetic damage occurred, since an increased level of respiratory-deficient mutants was frequently observed. Although such mutants would not be expected to materially affect fermentation behaviour, their presence suggests that freeze-drying can affect specific DNA sites, and this could be an important factor to bear in mind in the biotechnological use of constructed strains.

Changes in growth requirements have been reported (Kocková-Kratochvilová & Blagodatskaja, 1974) in strains of *Candida*. Again, in a study of the stability of the ade 1 and ade 2 loci of four strains of *Saccharomyces cerevisiae* following freeze-drying (freezing at −20 °C in 5% glycerol followed by vacuum drying) and drying (vacuum drying of cells suspended in distilled water and held on a Millipore filter), irreversible genetic changes were reported (Hieda & Ito, 1973). Using a single strain of baking yeast, Souzu (1973) reported phospholipid degradation in unprotected cells, resulting in some loss of viability.

Advantages and disadvantages. The long-term survival of yeasts that have been freeze-dried is an attractive advantage, since cultures so preserved can be stored without further attention until required for further use. This is particularly useful for service culture collections or industrial laboratories supplying material to other laboratories within the company. It is valuable for experimental purposes in providing an unlimited quantity of 'standard inocula', since variation between ampoules in any freeze-drying run is minimal.

However, the low level of survival obtained with most yeast strains suggests the need to establish that resistant sections of the original population have not been selectively preserved. The strain specificity of yeasts to freeze-drying damage prohibits the preservation of cultures composed deliberately of one or more strains, since the relative proportions are unlikely to remain constant. If mixed cultures are to be preserved by freeze-drying, it is generally more reliable to preserve the separate strains independently and to recombine them in the appropriate proportions after rehydration and individual propagation.

The stability of yeast characteristics confers substantial advantages over subculturing for long-term preservation, but indications of genetic damage observed suggest the need to determine the extent to which this

occurs, particularly in strains maintained for the conservation of properties controlled by single genes.

The initial capital investment for equipment may be high, as is the
requirement for skilled technical support and, in developing countries,
back-up for replacement of damaged equipment may be inadequate.

4.2.6   *Freezing in liquid nitrogen (cryopreservation)*

Introduction. Of all methods used for the preservation of biological material, cryopreservation is the one most widely applied and with
the greatest success. Although cells ranging from animal and plant cells
to microorganisms have been frozen and thawed with high survival
levels and good phenotypic stability, it has been necessary to determine
suitable protocols. Success can be affected by such intrinsic factors as cell
type and condition of the cells at the time of freezing, the latter
determined by growth phase and cultural conditions. Extrinsic factors
controlling cell survival include the cryoprotectant and its mode of
action, the rate of cooling and warming and the storage temperature.

As with freeze-drying, post-logarithmic cells survive better than
younger cells but, conversely, cells grown under aerobic conditions on a
shaker generally lead to higher viability figures, suggesting a requirement for a different membrane composition for protection against
freezing stresses. However, work in this area has been carried out with a
limited number of test strains and the conclusions may not be generally
applicable.

Different cryoprotectants have been used for the freezing of yeasts
and success has been recorded with 5, 10 and 20% glycerol, glycerol plus
dimethyl sulphoxide, 10% dimethyl sulphoxide, ethanol, methanol,
YM broth and 5 or 10% hydroxyethyl starch. The use of 5% glycerol has
been found to be applicable for a wide range of yeasts (Kirsop & Henry,
1984). The effectiveness of different cryoprotectants depends on their
molar concentration and the ease with which they penetrate the cells.

The rate at which cells are cooled and thawed is probably the most
important factor affecting cell survival, and different cell types respond
variably to different cooling rates. For red blood cells, for example, the
optimum cooling rate of unprotected cells is about $3000\,°C\;min^{-1}$,
whereas for yeast cells it is in the order of $10\,°C\,min^{-1}$ (Mazur, 1970). It is
thus not possible to obtain good recovery if yeast cells are immersed
directly into liquid nitrogen when the cooling rate is in the order of
$200\,°C\;min^{-1}$.

It is commonly held that damage to cells during freezing and thawing is caused by two principal factors. At fast cooling rates, ice crystal formation is a major factor of cell death, damaging cell membranes and other cellular components. At slower cooling rates, ice is formed outside the cells, thus increasing the concentration of solutes and leading to loss of water from the cells and subsequent shrinkage (Morris, 1981). Cells respond differently to these two factors and a cooling protocol can only be established by balancing the two stresses to reach conditions that lead to minimal cellular damage. Cooling rate experiments can be carried out using a controlled freezing bath, but this equipment is not always available, nor is it practicable to establish individual optimum procedures for large numbers of strains. The difficulty can be resolved in practice by using a 2-stage or interrupted cooling system, which cools cells to -20 or -30 °C and allows dehydration to take place over a period of 1–2 h. After an optimum amount of water has been removed from cells in this way, they may be immersed in liquid nitrogen at high cooling rates without further danger of ice crystal damage. With a wide range of yeasts, cooling to -30 °C and dehydrating for 2 h has proved successful.

It has been found possible to obtain some rough control of the freezing rate of cell suspensions by placing ampoules in plastic foam boxes which are placed in the vapour phase of liquid nitrogen until freezing has occurred. The ampoules are then transferred to the liquid nitrogen.

With all cell types a fast warming rate confers the highest survival levels, since rapid thawing of extracellular ice leads to the least exposure of cells to high osmotic conditions and subsequent damage.

The temperature of storage can affect long-term survival and, using liquid nitrogen, cells may be held either at -196 °C in the liquid phase, or at about -130 to -100 °C in the vapour phase. It is held that molecular activity in biological material can continue at temperatures above -139 °C and it follows that storage in the liquid phase is preferable. Care must be taken, however, to prevent liquid nitrogen seeping through the caps of polypropylene cryotubes, and this can be done either by sealing in polypropylene sleeves or using the polypropylene straw method described below.

Inoculation, storage and recovery (straw method). Practical details of this method are described in *Maintenance of Microorganisms* (see Kirsop, 1984). Polypropylene straws are cut into lengths (Fig. 4.4) and formed into mini cryotubes by holding them in forceps close to a Bunsen flame and allowing them to melt and form a firm seal (Fig. 4.5). After autoclaving, the straws are filled aseptically with a suspension of equal proportions of a post-logarithmic broth culture and 10% glycerol, care being taken not to overfill the straws or wet the sealing edge. The filled straws are sealed by melting the polypropylene as before and placing them into cryotubes, 5 or 6 to each tube. These are cooled in a freezer or cooling bath at $-30\,°C$ for 2 h to allow primary freezing and dehydration, after which they are transferred to the liquid nitrogen refrigerator (Fig. 4.6) at $-196\,°C$, care being taken to prevent warming between primary and secondary freezing. Steps must be taken to ensure the level of liquid nitrogen in the refrigerator is not allowed to fall below the level of the stored ampoules.

Thawing and recovery is achieved by removing a straw and placing it in a water bath at $35\,°C$. The straw is opened aseptically with scissors and the contents transferred to an appropriate growth medium using a Pasteur pipette.

During all manipulations involving the use of liquid nitrogen, safety precautions must be strictly observed. Face masks and gloves should always be worn and special care taken when thawing cultures, as any liquid nitrogen that has seeped into the ampoules may cause explosions as a result of rapid expansion on warming. It is for this reason that polypropylene containers should be used in preference to glass containers wherever possible. Additionally, refrigerators must be placed in

Fig. 4.4. Polypropylene straws for use in cryopreservation.

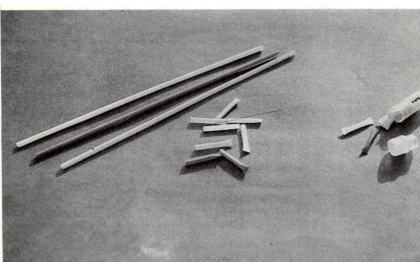

a well-ventilated room to prevent accidents caused to personnel by a build-up of nitrogen gas and depletion of oxygen in the atmosphere.

Survival and stability. The survival figures shown in Table 4.2 are for yeasts preserved by the standard method described in Section 4.2.6, without attempts to obtain optimum cooling conditions for individual strains. They show a substantial improvement over figures obtained for freeze-drying (see Table 4.1) and can be expected to be improved with

Fig. 4.5. Sealing polypropylene straw.

appropriate adjustments to procedures. Viable counts exceeding the inoculum level are sometimes obtained as the freezing process breaks up clumps and chains of cells, increasing the number of colony-forming units.

Good stability of phenotypic characters following cryopreservation has been reported (Wellman & Stewart, 1973; Hubalek & Kochová-Kratochvilová, 1978; Kirsop & Henry, 1984) and, with the high survival

Fig. 4.6. Storage of cryotubes in liquid nitrogen refrigerator.

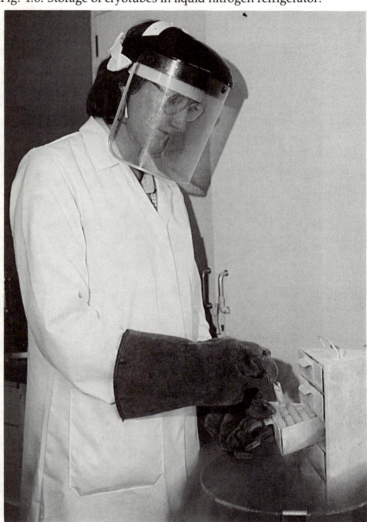

levels obtained, little genetic damage can be expected. Nevertheless genetic change has been reported in experiments in which cryoprotectants were not used (Calcott, Wood & Anderson, 1983), and it is clear that every effort must be made to establish optimum cryopreservation conditions for individual strains.

Advantages and disadvantages. There is no other method for yeast maintenance that yields the high survival levels and strain stability obtained with cryopreservation. However, the method requires a high initial capital outlay, and running costs are not negligible, particularly in hot countries where evaporation rates of liquid nitrogen will be high. Moreover, a reliable supply of liquid nitrogen must be assured, as failure of delivery could lead to serious losses. It is prudent to keep valuable cultures by more than one method, and most service culture collections adopt this principle.

A further disadvantage that must be accepted with this method is the inconvenience of regularly replenishing refrigerators even when laboratories or factories are closed for holiday periods. In addition, although the method is technically simple, staff must be trained to observe the safety precautions (see above) if accidents are to be avoided.

Table 4.2. *Survival levels of yeast genera following cryopreservation*

Genus	% Survival	No. of strains
Brettanomyces	29.2	4
Candida	70.4	25
Cryptococcus	85.4	3
Debaryomyces	54.8	3
Hanseniaspora	65	1
Hansenula	51.4	7
Kloeckera	34.2	2
Kluyveromyces	62.4	6
Lipomyces	68.7	7
Pichia	50.6	10
Rhodotorula	90.3	1
Saccharomyces	66.1	75
Schizosaccharomyces	103	2
Sporobolomyces	62	5
Trichosporon	85.6	2

From Kirsop & Henry (1984).

4.2.7   Freezing in mechanical freezers

Mechanical freezers operating at -20, -70, -80, -90 or $-135\,°C$ are available and may be used for the preservation of yeasts. As with preservation in liquid nitrogen, a cryoprotectant is necessary in order to obtain good survival levels at the lower storage temperatures; at the higher temperatures cryoprotectants are not always used. Little quantitative data are available for viability and stability of yeasts stored in refrigerators at these temperatures, although satisfactory survival levels for strains frozen in 10% glycerol have been reported (Mikata & Banno, 1986) after 2 years' storage. As has been discussed before (Section 4.2.6), molecular activity can occur at temperatures above $-139\,°C$ and use of higher temperatures may lead to less strain stability. For shorter-term storage, freezing at these temperatures may be acceptable, but the risk of losses due to mechanical breakdown must be accepted and back-up arrangements made.

The refrigerator operating at $-135\,°C$ (White & Wharton, 1984) shares with liquid nitrogen an acceptably low storage temperature and yet does not have the safety disadvantages described in Section 4.2.6. If it can be associated with a reliable back-up procedure, it is an attractive option, providing the substantial capital outlay can be justified.

4.3   Storage of genetically manipulated strains

Developments in biotechnology make it increasingly important to establish preservation systems for the storage of constructed strains. In research laboratories strains used in developmental work are generally stored in mechanical freezers at temperatures between -20 and $-90\,°C$, generally with varying levels of glycerol as a cryoprotectant. There is no documented information on the success of these methods, but it is generally agreed that from time to time, strains fail to behave in the expected manner.

Little is known about the effects of cooling on plasmid DNA; there is a need for research on this to establish protocols most likely to preserve material unchanged over prolonged storage periods. It is common practice in research laboratories to go back to parent strains when plasmids appear to be lost or genes fail to be expressed. This practice is not acceptable, however, in a culture collection committed to the supply of reliable material nor in an industrial situation in which strain stability is an essential criterion of success.

In the light of present knowledge, it would seem probable that storage in liquid nitrogen is likely to be preferable to storage at higher

Table 4.3. *Comparison of preservation systems*

	Subculturing	Drying	L-Drying	Freeze-drying	Cryopreservation
Survival level	fair	low	high	low	high
Shelf life	short	?	long	long	long
Strain stability	poor	poor	good	good	very good
Suitability for wide range of yeasts	good	?	good	poor	good
Convenience for distribution	poor	poor	very good	very good	poor
Level of technical expertise needed	low, but commitment needed	low	high	high	medium
Capital outlay	low	low	medium	high	high
Running costs	low	low	medium	medium	high
Safety to personnel	good microbiological practice required				additional care needed (see Section 4.2.6)

temperatures, but this judgement is based on data for wild-type strains, rather than on experimental evidence on genetically manipulated strains. The IFO collection (Chapter 1) reports that L-drying (Section 4.2.4) has been used successfully for the storage of genetic material. Kaneko, Mikata & Banno (1985) have shown that L-dried recombinant strains of *Saccharomyces cerevisiae* survive well and show little subsequent loss in viability following accelerated storage. There is some loss in maintenance of plasmids immediately following L-drying, but this does not continue during storage. Different recombinant strains vary in the extent to which plasmid loss occurs.

4.4   Summary and selection of methods

Table 4.3 provides a guide to the relative merits of subculturing, drying, L-drying, freeze-drying and cryopreservation with respect to factors that need to be considered in the selection of a preservation system. The major factor affecting the selection of a preservation system must be the degree to which the material to be preserved remains viable and genetically stable. The investment that can be justified for equipment and labour will depend on the importance of the cultures to be maintained and the time-scale for which storage is required. Additional criteria affecting the choice are the frequency with which cultures will be required and whether distribution services are provided. The existing levels of expertise, equipment and storage space available will be factors to be taken into account as will the numbers of strains to be maintained. Financial constraints will need to be considered and balanced against the value of the material to the user. Unique and valuable production strains warrant the use of the method most likely to ensure continuing survival and stability, whereas strains forming part of an earlier screening programme and only of potential interest to the future could well be maintained using a cheaper and less-secure system.

5

Identification and taxonomy

C. P. KURTZMAN

5.1 Introduction

Yeasts are a phylogenetically diverse group of fungi whose teleomorphs or sexual states are found among two major taxonomic classes, the Ascomycotina and the Basidiomycotina. Consequently, the term yeast is one of convenience that describes predominantly unicellular organisms whose means of vegetative cell division is either by budding or by fission. Furthermore, in contrast to higher fungi, the sexual state of a yeast is not enclosed in a fruiting body.

Microbiologists have studied yeast taxonomy for well over a century, but despite considerable progress, the task of developing an accurate system of classification is far from complete. The need for reliable identifications is readily apparent for proper treatment of infections, for understanding species interactions in nature, and for selection of appropriate organisms in food manufacture and industrial fermentations. If a reliable classification system can be developed that has its basis in phylogeny, then the scheme assumes added importance because it allows prediction of genetic similarity, including metabolic pathways and the mechanisms for controlling these pathways. Such a phylogenetic system will be of particular interest to the biotechnologist who may have need to combine diverse metabolic capabilities, and if these capabilities are from closely related species, their chance of integration and expression in a single organism is much greater. For example, evidence from protoplast fusion experiments suggests that stability of the fusants is greater when they are more closely related (Ferenczy, 1981). In the case of properties added through recombinant DNA technology, one would expect the expression of newly inserted gene sequences to be more efficient in organisms showing closer relatedness to the donor species.

The purpose of this chapter is to give those who are not yeast taxonomists a general overview of the methods for identifying yeasts and an assessment of the stability of the characters measured as perceived from genetic and molecular comparisons. The information given will permit the non-specialist a reasonable chance to identify isolates to genus, and the references given will allow placement at the species level. Also provided is a brief discussion of industrial processes mediated by yeasts with comments on the taxonomy of the species involved.

5.2   Standard identification methods

5.2.1   *General procedures*

Separation of yeasts into genera and occasionally into species is usually made through microscopic appearance of vegetative and sexual states. Consequently, these morphological forms are described below and again in Tables 5.1–5.3. Most observations can be made at around 400–500× with a standard brightfield microscope. Occasionally, an oil immersion lens (*c.* 1000× total magnification) is needed to detect ascospore ornamentation, and phase contrast can sometimes be helpful.

Physiological tests, also described in some detail in Section 5.2.4, represent the next step in identifying yeasts to species. Most taxonomists use these tests in conjunction with microscopic observations, but there are several schemes that allow complete identification of taxa nearly exclusively on the basis of fermentation and assimilation reactions. The main disadvantage to exclusive reliance on physiological tests is that any variability in test reactions, either due to procedures or to strain response, can result in totally erroneous identifications. The different systems available to the yeast worker are described in Section 5.2.5.

5.2.2   *Morphology of vegetative states*

The large majority of yeasts exhibit multilateral budding, a type of cell division in which buds appear over a wide area of the cell surface. This is in contrast to bipolar budding where buds appear only at the poles of the cell. One further variation is unipolar budding in which new buds emerge only from one pole of the cell. This type of budding characterises the genus *Malassezia*. Examples of the various types of

Table 5.1. *Ascomycetous yeast genera and their characteristics*[a]

| Genus | No. of known species | Appearance of | | | Other features |
		ascus	ascospores	vegetative cells	
Ambrosiozyma	4	Spheroidal or ovoidal, generally formed on hyphae, usually evanescent	Hat-shaped, 1–4 per ascus	Multilateral budding, pseudohyphae, true hyphae with dolipore septa	Species show slow or weak fermentation of sugars; frequently isolated from insects or insect tunnels in woody plants
Arthroascus	1	Swollen hyphal cells, evanescent	Spheroidal to subspheroidal with a circumfluent ledge, sometimes warty, 1–4 per ascus	Budding mostly multilateral, pseudohyphae, true hyphae with septa having a single micropore; hyphal cells break apart easily	No fermentation of sugars; isolated from soil and plant materials
Arxiozyma[b]	1	Spheroidal or ovoidal, arise from conversion of vegetative cells, persistent	Spheroidal to ovoidal, warty, 1–2 per ascus	Multilateral budding, pseudohyphae, no true hyphae	Ferments sugars; isolated primarily from poultry and other birds; previously classified as *Saccharomyces telluris*
Citeromyces	1	Mostly spheroidal, sometimes conjugated, persistent	Spheroidal and warty, 1–2 per ascus	Multilateral budding, lacks pseudohyphae or true hyphae	Strong fermentation of some sugars; isolated from sugar concentrates and tree fluxes

Table 5.1. (cont.)

| Genus | No. of known species | Appearance of | | | Other features |
		ascus	ascospores	vegetative cells	
Clavispora	1	Spheroidal to elongate, formed following conjugation of complementary mating types, evanescent	Clavate and finely warted, 1–2 per ascus	Multilateral budding, pseudohyphae, no true hyphae	Ferments sugars; primarily from human and animal sources
Coccidiascus	1	Elongate and banana-shaped to crescent-shaped	Spindle-shaped but in a helix; up to 8 per ascus	Multilateral budding, pseudohyphae and true hyphae unknown	The genus is known only from the tissue of *Drosophila* spp.
Cyniclomyces	1	Elongate, seldom evanescent	Generally elongate, 1–4 per ascus	Multilateral budding but predominantly near the poles of the cells; pseudohyphae but no true hyphae	Weakly ferments sugars; occurs in stomachs and faeces of rabbits
Debaryomyces	9	Spheroidal to ellipsoidal, mother-bud conjugation, occasionally conjugation between independent cells, persistent	Generally spheroidal with warted surface, one species with spiral ridges and warts, 1–4 per ascus	Multilateral budding, occasionally pseudohyphae, no true hyphae	Weak to occasionally strong fermentation of sugars; common to soil, foods, salted products, many plant products, insects, sea-water, clinical specimens

Genus	No.	Ascospores	Vegetative reproduction	Physiology and habitat
Dekkera	2	Spheroidal to elongate, unconjugated, evanescent	Hat-shaped or spheroidal with tangential brims, 1–4 per ascus	Multilateral budding, cells frequently ogival at one end, pseudohyphae, no true hyphae	Ferments sugars and fermentation is stimulated by oxygen; isolated as a contaminant of beer and wine
Guilliermondella	1	Formed as an elongated cell following conjugation between hyphal cells or cells of adjacent hyphae, evanescent	Spheroidal, ovoidal, reniform or falcate with two appendages on opposite sides of the ascospore, 1–4 per ascus	Budding predominantly near the poles of the cells, pseudohyphae and true hyphae with septa having plasmodesmata	Ferments glucose; isolated from tanning liquor
Hanseniaspora	6	Ovoidal to elongate, unconjugated, evanescent for species with hat-shaped ascospores	Hat- to helmet-shaped or spheroidal and smooth or warty with or without a ledge, 1–4 per ascus	Apiculate (lemon-shaped) cells showing budding, may form pseudohyphae but not true hyphae	Ferments sugars; primarily from soil and fruits but also from other plant materials and from clinical sources
Issatchenkia	4	Ovoidal to elongate, unconjugated unless formed by pairing of complementary mating types, persistent	Spheroidal and finely to roughly warty, 1–4 per ascus	Multilateral budding, pseudohyphae, no true hyphae	Ferments sugars; from soil, fruits, berries, wine, fluxes of trees, *Drosophila*, sea-water and clinical specimens
Kluyveromyces	12	Spheroidal to elongate, generally unconjugated, evanescent	Spheroidal to reniform, smooth, 1–c. 100 per ascus	Multilateral budding, may form pseudohyphae but not true hyphae	Ferments sugars; from soil, water, fruit and other plant materials, tree fluxes, dairy products, *Drosophila*, clinical specimens

Table 5.1. (cont.)

Genus	No. of known species	Appearance of			Other features
		ascus	ascospores	vegetative cells	
Lipomyces	4	Ellipsoidal to elongate, frequently conjugated, generally evanescent	Spheroidal to elongate, with warts or ridges, pigmented amber to brown, 1–c. 30 per ascus	Multilateral budding, may form rudimentary pseudohyphae but no true hyphae	No fermentation of sugars; colonies mucoid; isolated from soil
Lodderomyces	2	Spheroidal to ellipsoidal, unconjugated, persistent	Oblong with blunt ends and somewhat tapered, 1–2 per ascus	Multilateral budding, pseudohyphae, no true hyphae	Ferments sugars; utilises higher alkanes; from soil, orange juice
Metschnikowia	6	Clavate to sphaero- or ellipsopedunculate, unconjugated, persistent	Elongated and generally dart-like, 1–2 per ascus	Multilateral budding, pseudohyphae, no true hyphae	Ferments sugars; from invertebrates, sea-water, fresh water, flowers, *Drosophila*
Nadsonia	3	Elongate and formed by mother–bud conjugation; depending on the species, the mother cell or an opposite bud becomes the ascus, persistent	Spheroidal, roughened, 1–2 per ascus	Bipolar budding, neither pseudohyphae nor true hyphae	May ferment sugars; from soil and the slime fluxes of trees
Nematospora	1	Elongate, evanescent	Elongated and spindle-shaped with a whip-like terminal appendage	Predominantly multilateral budding, some cells with unusual shapes, pseudo- and true hyphae	Ferments sugars; causes disease of hazelnuts, cotton bolls, various beans

Pachysolen	1	Hemispheroidal and formed in the flared end of a heavy-walled tube-like ascophore, evanescent	Hemispheroidal with a narrow ledge at the base, 1–4 per ascus	Multilateral budding, pseudohyphae, no true hyphae	Ferments sugars including xylose; from tanning liquors
Pachytichospora	1	Ellipsoidal, unconjugated, persistent	Spheroidal to ellipsoidal with thick walls, 1–2 per ascus	Multilateral budding, rudimentary pseudohyphae, no true hyphae	Ferments sugars; from soil, animal caecum
Pichia	90	Spheroidal, ellipsoidal, conjugated and unconjugated, persistent and evanescent	Spheroidal, hat-shaped, roughened, spheroidal with a ledge, usually 1–4 per ascus	Multilateral budding, some species with pseudohyphae and true hyphae which have a single septal micropore	Some species ferment sugars; from soil, trees, fruits, water, insects, clinical specimens; hat-spored species of *Hansenula* included in the genus
Saccharomyces	9	Spheroidal to ellipsoidal, unconjugated, persistent	Spheroidal to slightly ellipsoidal, smooth, 1–4 per ascus	Multilateral budding, pseudohyphae, no true hyphae	Species generally give a vigorous fermentation of sugars; from soil, fruits, foods, beverages, rarely in clinical specimens
Saccharomycodes	1	Ellipsoidal to elongated, unconjugated, persistent	Spheroidal with a fine subequatorial ledge, 1–4 per ascus	Apiculate cells showing bipolar budding, pseudohyphae but no true hyphae	Ferments sugars; from slime fluxes of trees

Table 5.1. (*cont.*)

Genus	No. of known species	Appearance of			vegetative cells	Other features
		ascus	ascospores			
Saccharomycopsis	6	Spheroidal to ellipsoidal, free or attached to hyphae, unconjugated, evanescent or persistent	Hat-shaped, spheroidal with 1 or 2 ledges or ellipsoidal with a ledge and roughened, 1–4 per ascus		Multilateral budding, pseudohyphae, true hyphae with septa having plasmodesmata	Slowly or weakly ferment sugars; from trees, pollen, fruit, insects, various starchy foods
Schizosaccharomyces	4	Elongated, conjugated or unconjugated, persistent or evanescent	Spheroidal to ellipsoidal, smooth, generally 4–8 per ascus		Cell division by fission, true hyphae may be formed	Species generally vigorously ferment sugars; from fruits and their juices, wines, high sugar substrates
Schwanniomyces	1	Spheroidal to ellipsoidal, mother-bud conjugation, rarely conjugation between independent cells, persistent	Spheroidal, roughened and with a distinct equatorial ring, 1–2 per ascus		Multilateral budding, pseudohyphae rudimentary and seldom formed, no true hyphae	Sugars fermented; from soil
Sporopachydermia	2	Spheroidal to elongate, conjugated or unconjugated, evanescent	Spheroidal to ellipsoidal, smooth, 1–4 per ascus		Multilateral budding, occasional rudimentary pseudohyphae, no true hyphae	No fermentation of sugars; produces an offensive odour; from sea-water, fresh water, waste water, rots in cacti, clinical specimens

Genus					
Stephanoascus[c]	2	Spheroidal with an apical cap cell, asci form after conjugation of hyphal cells from complementary mating types, persistent	Initially hat-shaped and maturing to hemispheroidal with the ledge becoming a thickened wall, 1–4 per ascus	Multilateral budding, pseudohyphae, true hyphae with septa having plasmodesmata	No fermentation of sugars; from soil, human and animal sources
Torulaspora	3	Spheroidal to ellipsoidal and showing mother cell–bud conjugation which frequently leaves the appearance of a tapered outgrowth, conjugation between cells also occurs, persistent	Spheroidal and usually roughened, 1–4 per ascus	Multilateral budding, rudimentary pseudohyphae, no true hyphae	Sugars fermented; from soil, fruits, fruit juices, alcoholic beverages, foods high in sugar or salt, animals, clinical specimens
Waltomyces[c]	1	Ellipsoidal to elongate, frequently conjugated, generally evanescent	Spheroidal to elongate, smooth, pigmented amber to brown, 1–20 per ascus	Multilateral budding, pseudohyphae and true hyphae absent	No fermentation of sugars; colonies mucoid; isolated from soil; this genus was split from *Lipomyces* because the species has smooth ascospores and coenzyme Q–10 in contrast to other species of *Lipomyces* which have ornamented ascospores and coenzyme Q–9

Table 5.1. (cont.)

Genus	No. of known species	Appearance of			
		ascus	ascospores	vegetative cells	Other features
Wickerhamia	1	Ellipsoidal, unconjugated, evanescent	Shape is unusual and somewhat resembles a baseball cap, 1–2 per ascus	Bipolar budding, no pseudohyphae or true hyphae	Sugars are fermented; the single isolate was from squirrel dung
Wickerhamiella	1	Spheroidal and conjugated, at maturity, one end dehisces ejecting the ascospore	Oblong with obtuse ends and roughened, usually 1 per ascus	Multilateral budding, no pseudohyphae or true hyphae	Sugars are not fermented; from wine vat, sugar factory
Williopsis	1	Spheroidal to ellipsoidal, mother cell–bud and cell–cell conjugation, usually evanescent	Spheroidal to ellipsoidal with an equatorial ring, 1–4 per ascus	Multilateral budding, pseudohyphae, no true hyphae	Sugars are fermented; from soil, water, tree frass, slime fluxes, animal dung
Wingea	1	Spheroidal and conjugated or with an elongated tube, persistent	Lens-shaped and light brown in colour, 1–4 per ascus	Multilateral budding, no pseudo- or true hyphae	Sugars are fermented; from soil and larval feeding areas
Yarrowia[a]	1	Spheroidal or ellipsoidal, usually unconjugated, evanescent	Shape is spheroidal and roughened, hat-shaped, crescentiform or saucer-like, depending on mating types, 1–4 per ascus	Multilateral budding, pseudohyphae and true hyphae with a single central pore	Sugars are not fermented; from soil, processing wastes, lipoidal and proteinaceous materials, animal and clinical specimens

Zygoascus[e]	1	Spheroidal to ellipsoidal and formed on the bridge between two conjugating hyphal cells of opposite mating type, persistent	Hemispheroidal to hat-shaped, 1–4 per ascus	Multilateral budding, pseudohyphae and true hyphae with a single septal micropore	Sugars are fermented; from grape must, organic industrial waste, mastitic cow
Zygosaccharomyces	8	Generally comprised of 2 conjugated cells each containing 1 or more ascospores, persistent	Spheroidal to ellipsoidal, 1–4 per ascus	Multilateral budding, pseudohyphae, no true hyphae	Sugars are fermented; from wine, various foods, fruit, trees, *Drosophila*

[a] From Kreger-van Rij (1984) except where noted.
[b] van der Walt & Yarrow (1984b).
[c] Yamada & Nakase (1985).
[d] van der Walt & von Arx (1980).
[e] Smith (1986).

Table 5.2. *Basidiomycetous yeast genera and their characteristics*[a]

| Genus | No. of known species | Appearance of | | Other features |
		sexual state	vegetative cells	
Chionosphaera	1	Uninucleate budding cells of opposite mating type conjugate giving rise to dikaryotic hyphae which form a small fruiting body; basidia on the fruiting body bear basidiospores which reproduce by budding	Budding cells and occasional pseudohyphae, true hyphae form only as part of the sexual cycle	Except for the presence of a fruiting body, this genus seems to have a life cycle similar to *Filobasidiella*; neither pigments nor teleospores are formed; from dead tree limbs
Filobasidiella	1	Budding cells of opposite mating types conjugate giving rise to dikaryotic hyphae with clamp connections; non-septate basidia arise from these hyphae to produce basidiospores which form in chains at four foci on the tip of the basidium; homothallic self-sporulating strains are also known	Budding cells; pseudohyphae do not form from haploid cells	Teleomorphic state of *Cryptococcus neoformans*; neither pigments nor teleospores are formed; from soil, pigeon excretia
Filobasidium	3	Budding cells of opposite mating types conjugate leading to the formation of dikaryotic hyphae with clamp connections; basidia arise from these hyphae and form terminal basidiospores; self-sporulating strains are also known	Budding cells, pseudo- and true hyphae may be formed by some mating types	Sugars are fermented by one species; cultures are not pigmented; teleospores are not formed; from alcoholic beverages, fermented food, clinical specimens

Leucosporidium	6	Life cycles varied; budding cells of opposite mating types conjugate to form dikaryotic hyphae with clamp connections which gives rise to teleospores; teleospores germinate to form septate or non-septate basidia which form basidiospores; self-sporulating strains show no prior conjugation and have hyphae lacking clamp connections	Budding cells; pseudo- and true hyphae formed by some strains prior to conjugation	Most species ferment sugars; pigments are not formed; from sea-water, Antarctic soil
Rhodosporidium	9	Life cycles are varied and generally similar to those described for *Leucosporidium*	Budding cells; pseudo- and true hyphae formed by some strains prior to conjugation	Sugars are not fermented; cultures are shades of yellow, orange and red due to the presence of carotenoid pigments; from sea-water, fresh water, soil, plants, clinical specimens
Sporidiobolus	4	Life cycles are varied and rather similar to those described for *Leucosporidium*	Budding cells; pseudo- and true hyphae formed by some strains prior to conjugation; *Sporidiobolus* and its anamorph *Sporobolomyces* are characterised by the formation of ballistospores	Sugars are not fermented; cultures are shades of pink or red due to the presence of carotenoid pigments; from plant leaves, soil, water, air, clinical specimens

Table 5.2. (cont.)

| Genus | No. of known species | Appearance of | | Other features |
		sexual state	vegetative cells	
Sterigmatospori-dium[b]	1	Budding cells of opposite mating types conjugate and form dikaryotic hyphae which bear teleospores; these germinate to form non-septate basidia which produce basidiospores; self-sporulating isolates are also known	Buds are formed on cellular extensions termed sterigmata; pseudo- and true hyphae are also formed	Sugars are not fermented; cultures are not pigmented; from water-soaked wood

[a] From Kreger-van Rij (1984) except where noted.
[b] Kraepelin & Schulze (1982).

Table 5.3. *Anamorphic (imperfect) yeast genera and their characteristics*[a]

| Genus | No. of known species | Appearance of | | Other features |
		vegetative cells	expected teleomorph	
Aciculoconidium	1	Budding ovoid and ellipsoidal cells in addition to the diagnostic needle-shaped conidia; true hyphae are also formed	Ascomycete on the basis of the negative DBB test[c]	Sugars are weakly fermented; from *Drosophila*
Brettanomyces	7	Multilateral budding, cells frequently ogival at one end, pseudohyphae are formed but not true hyphae	The genus *Dekkera* for which some sexual states have already been formed	See *Dekkera* in Table 5.1
Bullera	6	Budding cells, pseudo- and true hyphae; in addition, ballistospores are formed	Basidiomycete on the basis of a positive DBB test and the presence of ballistospores	Sugars are not fermented; carotenoid pigments lacking; from frass in various trees, air, plant debris
Candida	165	Multilateral budding, pseudo- and true hyphae	Many genera of ascomycetes are represented in this phylogenetically diverse form genus; in addition, a few of the species show affinities with some of the genera of basidiomycetes	The characteristics are those of teleomorphic genera showing multilateral budding

Table 5.3. (*cont.*)

| Genus | No. of known species | Appearance of | | Other features |
		vegetative cells	expected teleomorph	
Cryptococcus	24	Budding cells and occasionally pseudohyphae	On the basis of the DBB test, nearly all species are expected to be basidiomycetes	Sugars are not fermented, inositol serves as a sole source of carbon, carotenoid pigments usually lacking; from air, water, foods, various plant material, insects, animals, humans
Eeniella[b]	1	Bipolar budding and occasionally rudimentary pseudohyphae	Negative DBB test indicating the genus to be ascomycetous	Sugars are fermented; from beer; this genus is unusual because it shows bipolar budding but has many characteristics of *Brettanomyces* including production of acetic acid and short life in culture
Fellomyces	5	Budding from the tips of sterigmata with the point of separation immediately adjacent to the bud; mother cells usually have several sterigmata; pseudo- and true hyphae may be present	Vegetative growth is similar to *Sterigmatosporidium* and this or a similar basidiomycetous genus probably represents the teleomorphic state	Sugars are not fermented; from air, sea-water, cheese, insects, animals, humans
Kloeckera	1	Apiculate (lemon-shaped) cells showing bipolar budding, may form pseudohyphae but not true hyphae	All former species of *Kloeckera* have had *Hanseniaspora* as their teleomorph	See *Hanseniaspora* in Table 5.1
Malassezia	2	Cells bud from only one pole; true hyphae may also be formed	On the basis of the DBB test, the species are expected to be basidiomycetes	Found on the skin of humans and animals

Oosporidium	1	Multilateral budding and occasionally true septa between buds; pseudohyphae may also be formed	On the basis of the DBB test, the species appears to be an ascomycete	Sugars are not fermented; pink or orange-yellow non-carotenoid pigments are present; from slime fluxes of trees
Phaffia	1	Budding cells, chlamydospores, and rudimentary pseudohyphae	The positive DBB test indicates the genus to be basidiomycetous	Weak fermentation of sugars; carotenoid pigments impart a salmon-red colour to cultures; from slime fluxes of various trees
Rhodotorula	8	Multilateral budding, pseudo- and true hyphae may be formed	Sexual states found in *Rhodotorula* have been basidiomycetous and assigned to the genus *Rhodosporidium*	Sugars are not fermented; carotenoid pigments give cultures a red, orange, or yellow colour; from air, water, soil, plant surfaces and debris, insects, animals, humans
Sarcinosporon	1	Budding cells that separate or remain in chains and sarcina-like aggregates of cells formed by septation in different planes; the cells may also contain vegetative endospores; true hyphae are also formed	Positive for the DBB test which indicates the genus to be basidiomycetous	Sugars are not fermented; the single strain was from a skin lesion
Schizoblastosporion	1	Budding is primarily, but not exclusively, bipolar; pseudohyphae are rudimentary or absent; no true hyphae are formed	Negative for the DBB test which indicates the genus to be ascomycetous	Sugars are not fermented; from soil, stomach contents
Sporobolomyces	7	Budding cells; some strains also form pseudo- and true hyphae; one of the major diagnostic characteristics is the formation of ballistospores	Sexual states found so far have been basidiomycetous and all were assigned to *Sporidiobolus*	Sugars are not fermented; cultures are shades of pink or red due to the presence of carotenoid pigments; from soil, water, air, plant leaves, clinical specimens

Table 5.3. (cont.)

Genus	No. of known species	Appearance of		Other features
		vegetative cells	expected teleomorph	
Sterigmatomyces	1	Budding from the tips of sterigmata with separation of cells at the midpoint of each sterigma, the main distinction from *Fellomyces*; pseudo- and true hyphae not formed	Strains are DBB positive and therefore basidiomycetous	Sugars are not fermented; from air, sea-water, clinical specimens
Sympodiomyces	1	Budding cells are present; in addition, conidiophores arise from some cells and produce terminal conidia; successive conidiation leaves a zig-zag sympodial-type pattern of scars; true hyphae are formed	Strains are DBB negative and considered ascomycetous	Sugars are not fermented; from sea-water
Trichosporon	10	Budding cells as well as arthrospores that are formed by a fission process; pseudo- and true hyphae are present; asexual endospores may also be formed	On the basis of the DBB test, both ascomycetous and basidiomycetous species are assigned to the genus	Sugars may or may not be fermented; from air, water, soil, foods, plant material, humans, animals
Trigonopsis	1	Multilateral budding with cells either ellipsoidal or triangular in outline; pseudo- and true hyphae are not produced	Strains are DBB negative and considered ascomycetous	Sugars are not fermented; from beer and grape must

[a] From Kreger-van Rij (1984) except where noted.
[b] Smith, Batenburg-van Vegte & Scheffers (1981).
[c] See p. 122.

budding are illustrated in Fig. 5.1. *Schizosaccharomyces* differs from other yeasts because vegetative division is exclusively through fission, a process by which a wall develops near the centre of the cell separating it in two. In contrast to budding, cells dividing by fission show little or no constriction at the site of the new wall (Fig. 5.1). One further variation concerns species of the genus *Trichosporon* which show both budding and fission. In only a few taxa, such as *Trigonopsis variabilis*, is cell shape sufficiently distinct to allow identification to species.

Besides the formation of single cells through budding or fission, certain species also develop hyphae and pseudohyphae. True hyphae are characterised by lack of constriction at the crosswalls, while pseudo-hyphal cells, although generally elongated, are formed by budding and show a constriction at their points of attachment to other cells (Fig. 5.2). Formation of hyphae and pseudohyphae is frequently stimulated by reduced oxygen. This can be effected on a Petri plate by placing a

Fig. 5.1. Types of budding found among the yeasts: (a) multilateral budding; (b) bipolar budding; (c) unipolar budding; (d) fission.

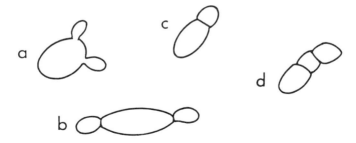

Fig. 5.2. Hyphal growth from yeasts: (a) pseudohyphae; (b) true hyphae.

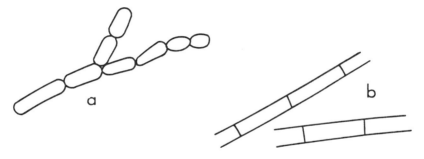

coverglass over the agar at the inoculation point. In addition to reducing oxygen, the coverglass facilitates microscopic observation of the developing growth.

5.2.3   Morphology of sexual states

Among the ascomycetous yeasts, taxonomic significance is assigned to whether the species has a homothallic or heterothallic mating system; to the type of conjugation, if any, before ascospore formation; to shape, surface topography, and number of ascospores; and to whether or not the ascus is evanescent. Heterothallic species are sometimes isolated from nature as asporogenous haploid mating types which then must be mixed with a complementary mating type before ascosporulation can occur. Homothallic yeasts produce no mating types, as such, but they may show conjugation prior to ascus formation. Conjugation can be between independent cells, or a cell and its bud may conjugate (Fig. 5.3). This latter configuration is termed mother–daughter conjugation. Both types of conjugation are found in sporulating cultures of many homothallic species.

Ascospores show considerable variety in shape and surface ornamentation. One of the most common forms is similar to a derby hat. Other shapes are spheroidal, ellipsoidal, saturn-shaped, dart-like and elongated with a terminal whip-like appendage (Fig. 5.4). Some of the spores also show surface warts and ridges. Further discussion of spore shape is included with generic descriptions (Table 5.1).

The basidiospore is the product of the sexual cycle among the basidiomycetous yeasts. In contrast to many ascospores, these spores are generally unornamented ellipsoidal or elongated cells, but the events leading to their biogenesis on the surface of basidia are frequently

Fig. 5.3. Forms of conjugation found among yeast species: (a) conjugation between independent cells; (b) mother–daughter conjugation. The bud serving as conjugant generally has a cell wall that appears thinner than that of the other buds.

complex. Depending on the genus, basidia may form from thick-walled teleospores or arise directly from hyphae (Fig 5.5). Species may be homothallic or heterothallic and, of the latter, the mating system may be unifactorial or bifactorial, that is, control of the mating system is by either one pair or two pairs of alleles. Among ascomycetes, heterothallic mating systems are unifactorial.

Fig. 5.4. Representative ascospores found among the yeasts: (a) hat-shaped, *Pichia anomala*; (b) spheroidal and smooth, *Saccharomyces cerevisiae*; (c) saturn-shaped and smooth, *Williopsis saturnus*; (d) spheroidal and roughened, *Debaryomyces hansenii*; (e) saturn-shaped and roughened, *Schwanniomyces occidentalis*; (f) elongated with terminal appendage, *Nematospora coryli*; (g) dart-like, *Metschnikowia reukaufii*.

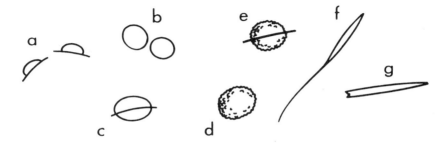

Fig. 5.5. Teleomorphs of basidiomycetous yeasts: (a) germinating teleospore with septate metabasidium bearing basidiospores, *Rhodosporidium toruloides*; (b) non-teleospore species with bulbous basidium bearing basidiospores, *Filobasidiella neoformans*.

5.2.4   *Physiological tests*

As has been discussed, the initial separation of yeasts into genera, with a few exceptions, is made from microscopic observations. However, recognition of species is frequently based on fermentation of, and growth responses to, a variety of carbon compounds.

Yeasts vary in their ability to ferment sugars as measured by the production of carbon dioxide. Some species give a vigorous fermentation while others show only a weak fermentation or no fermentation at all. While there are a variety of ways for measuring fermentation, yeast taxonomists have come to rely on Durham tubes because of their simplicity of preparation and ease of reading. The method of Wickerham (1951), as detailed in *The Yeasts, A Taxonomic Study* (Kreger-van Rij, 1984), is generally used, and the medium consists of a yeast-extract–peptone solution to which are added the test sugars, generally at a concentration of 2%. Inocula are heavy, and carbon dioxide initially produced from respiration serves to keep the system at a very low oxygen level. Fermentation is assessed by estimating the percent gas trapped in the inverted insert tube. Because readings are initially taken every day or two, a rough estimate of fermentation rate is also obtained. The sugars usually used for fermentation tests are D-glucose, D-galactose, sucrose, maltose, lactose, raffinose and trehalose. Occasionally, melibiose, D-xylose, starch and inulin are included for comparison. The ability of each genus to ferment sugars is given in Tables 5.1–5.3. Specific information for each species may be found in *The Yeasts, A Taxonomic Study* (Kreger-van Rij, 1984) and in *Yeasts: Characteristics and Identification* (Barnett, Payne & Yarrow, 1983). While it was initially thought that each species had a more or less immutable fermentation pattern, we now realise that this is not true and the reasons will be discussed in a later section.

The sugars and other carbon compounds used in assimilation tests, that is, the tests to measure whether these compounds support growth under aerobic conditions, are listed in Table 5.4. The tests are carried out in tubes of a synthetic liquid basal medium to which are added the various carbon sources, usually at the equivalent in carbon of 5 g of glucose per litre. The procedures and test compounds are those described by Wickerham (1951) and detailed methodology is given in *The Yeasts, A Taxonomic Study* (Kreger-van Rij, 1984). The inoculated tubes may be incubated on a shaker or held stationary, the latter method generally giving a slower growth response. Readings usually are taken at 1, 2, 3 and 4 weeks. Assimilation patterns for the various species may also be found in *The Yeasts, A Taxonomic Study* (Kreger-van Rij, 1984) and in *Yeasts: Characteristics and Identification* (Barnett *et al.*, 1983).

Test kits to measure growth by the auxanographic method are available from several commercial suppliers and are designed for rapid identification of medically important yeasts (Bowman & Ahearn, 1976; Baker *et al.*, 1981). Because the spectrum of carbon compounds in these kits is limited, they have less utility for identification of non-medical yeasts (see Section 5.2.5).

Other growth tests include response to nitrate, absence of vitamins, osmotic media, cycloheximide and various temperatures. Because the purpose of the tests is simply convenient and reliable discrimination of taxa in the laboratory, other compounds and growth conditions fitting these criteria are also suitable.

One of the major questions confronting those working with anamorphic yeasts is whether their strains are ascomycetes or basidiomycetes. Ultrastructure of the cell wall provides a reliable answer to this question (Kreger-van Rij & Veenhuis, 1971), but the procedures for transmission electron microscopy are not amenable to rapid and routine use in most laboratories concerned with yeast identification. Van der Walt & Hopsu-Havu (1976) reported that a simple chemical test could distinguish ascomycetes from basidiomycetes. When the dye Diazonium Blue B (DBB) reacts with cultures of basidiomycetes, the cell walls show an intense dark red colour. This colour does not occur with ascomycetes. Hagler & Ahearn (1981) introduced some modifications to the protocol for the DBB test.

Table 5.4. *Compounds typically used in assimilation and fermentation tests for the identification of yeasts*

Hexoses: D-glucose, D-galactose, L-rhamnose, L-sorbose
Pentoses: D-xylose, D-ribose, L-arabinose, D-arabinose
Disaccharides: sucrose, maltose, cellobiose, trehalose, lactose, melibiose
Trisaccharides: raffinose, melezitose
Polysaccharides: soluble starch, inulin
Alcohols: erythritol, ribitol (adonitol), D-mannitol, inositol, methanol, ethanol, glycerol, galactitol (dulcitol), D-glucitol (sorbitol)
Organic acids: succinic acid, citric acid, DL-lactic acid, malic acid, gluconic acid, glucuronic acid, 2-ketogluconate, 5-ketogluconate
Glycosides: α-methyl-D-glucoside, arbutin, salicin
Other compounds: glucono-δ-lactone, D-glucosamine hydrochloride, decane, hexadecane

5.2.5   *Identification systems for the general yeast worker*

Definitive identification. It is clear from the preceding sections that yeast identification is not simple, and if a definitive identification is required for publication, regulatory or legal purposes, it may be wise to obtain confirmation from an expert. Many specialists in yeast identification work in culture collections, and the centres that provide an identification service are listed in Chapter 1, Table 1.4. To obtain an identification, an examination of the morphological and physiological characteristics described above is carried out and reference is then made to the species descriptions in the standard taxonomic treatises. Keys to the different genera are provided, and their use leads to a species name. A comparison of the characteristics of the unknown yeast with those of the suspected taxon lead to a judgement as to whether the identification is sound. Experience is required to make a reliable assessment, since although some variation of phenotypic characters is known to occur, certain characteristics are considered more reliably discriminatory. When uncertainty remains, recourse may have to be made to the chemotaxonomic methods described below.

An attempt to simplify identification procedures has been made by Barnett *et al.* (1983), who used a computer to prepare a number of different keys based predominantly on physiological tests, so reducing the need for recognition of morphological features that may be difficult to detect without practice. The keys have recently been made available on disk (Cambridge University Press, 1985) and enable the selection of the species giving the same test results as the unknown yeast. The disadvantage of the system is that the large keys employed may lead to erroneous results in cases where the unknown yeast behaves atypically in a single test, or where errors in test procedures occur. The disk may be used to search for a single species with individual characteristics, or for a number of species with general properties in common. Strain variation occurs in some tests and although this is accommodated to some extent in the program, searches for microorganisms with specific properties can better be carried out on databases containing individual strain characteristics (see Chapter 2).

Recognising the shortcomings of dependence on keys, a probabilistic system has recently been developed (Kirsop *et al.*, 1986), based on that developed by Lapage *et al.* (1973) for enteric bacteria. A probability matrix has been prepared based on the response of over 400 species to 46 morphological and physiological tests, selected for reproducibility, ease of execution, independence and application to a wide range of species.

Unknown yeasts are examined by the tests (Table 5.5) and results are recorded and entered into the computer. The program compares the results with each record in the matrix, and the probability of the unknown yeast belonging to each taxon in the matrix is determined. The probabilities are multiplied to give a likelihood figure. The likelihoods are normalised to give an identification score, and the taxa with the highest scores are listed. The identification is thus based on the response of the unknown yeast to all tests, and an atypical test response is included in the probability calculation. It follows that if the investigator fails to detect ascospores, for example, the most probable identification will still be obtained, although the score may be slightly lowered. The system (COMPASS) is available on-line to those registered with the National Collection of Yeast Cultures (Chapter 1, Section 1.2.1) and equipped with access to computer networks. A typical COMPASS printout is shown in Table 5.6.

Presumptive identification. For situations requiring only an indication of the identity of yeasts, selected tests may be used to prepare specialised keys. For such systems to be successful, knowledge must exist of the genera and species likely to be encountered. In studies of the yeast flora of dairy products, for example, keys to species able to utilise lactose can be prepared, using the data published in the yeast taxonomy treatises.

Table 5.5. *Tests used in the COMPASS yeast identification system*

Test	Information obtained
Growth in YM Broth	Method of cell division
	Pellicle formation
Growth on corn meal agar	Vegetative growth characteristics
	Pseudomycelium
	True mycelium
	Arthrospores
	Ballistospores
Growth on potassium acetate agar	Ascospore formation
Fermentation of sugars	Ability to ferment 5 sugars
Aerobic growth in carbon sources	Ability to grow aerobically in 31 sources of carbohydrate
Aerobic growth in nitrogen sources	Ability to grow aerobically in 2 sources of nitrogen
Growth in vitamin-free medium	Ability to grow without added vitamins
Urease activity	Presence of urease activity

Table 5.6. *A typical COMPASS identification printout*

Name: C. BOIDINII

Record number: 1436

Identification score: 0.9999868   Likelihood: 0.3149113E−06

This is a probabilistic record

Character	Unknown	Identification	Probability
IDENTIFICATION			
Pellicle	1	1	0.9900000
Budding	2	2	0.9900000
Fission	0	0	0.9900000
Pseudomycelium	1	1	0.9900000
True mycelium	0	0	0.9900000
Ballistospores	0	0	0.9900000
Arthrospores	0	0	0.9900000
Endospores	0	0	0.9900000
Chlamydospores	0	0	0.9900000
Ascospores	0	0	0.9900000
Teliospores	0	0	0.9900000
F Glucose	2	2	0.9900000
F Galactose	0	0	0.9900000
F Sucrose	0	0	0.9900000
F Maltose	0	0	0.9900000
F Lactose	0	0	0.9900000
A1 Glucose	2	2	0.9900000
A2 Galactose	0	0	0.9900000

* * * * * * * * * * * * * The following result is unexpected   * * * * * * * * * * * * *

A3 Sorbose	2	0	0.1000000E−01
A4 Sucrose	0	0	0.9900000
A5 Maltose	0	0	0.9900000
A6 Cellobiose	0	0	0.9900000
A7 Trehalose	0	0	0.9900000
A8 Lactose	0	0	0.9900000
A9 Melibiose	0	0	0.9900000
A10 Raffinose	0	0	0.9900000
A11 Melezitose	0	0	0.9900000
A12 Inulin	0	0	0.9900000
A13 Soluble Starch	0	0	0.9900000
A14 Xylose	2	2	0.9900000
A15 L-Arabinose	0	x	0.5000000
A16 D-Arabinose	0	0	0.9900000
A17 Ribose	2	2	0.9900000
A18 Rhamnose	0	0	0.9900000
A19 Ethanol	2	2	0.9900000
A20 Glycerol	2	2	0.9900000

Table 5.6. (*cont.*)

Character	Unknown	Identification	Probability
IDENTIFICATION			
- - - - - - - - - - - - - -			
A21 Erythritol	2	2	, 0.9900000
A22 Ribitol	2	2	0.9900000
A23 Galactitol	0	0	0.9900000
A24 Mannitol	2	2	0.9900000
A25 Sorbitol	2	2	0.9900000
A26 a-M-D-Glucoside	0	0	0.9900000
A27 Salicin	0	0	0.9900000
A28 Lactic acid	2	1	0.9900000
A29 Succinic acid	0	0	0.9900000
A30 Citric acid	0	0	0.9900000
A31 Inositol	0	0	0.9900000
KNO3	2	2	0.9900000
Ethylamine	0	U	$0.9999990E-02$

* * * * * * * * * * * * * The following result is unexpected * * * * * * * * * * * * *

Vitamin free	0	1	$0.9999990E-02$

Systems of this kind are valuable for screening and making preliminary groupings of large numbers of isolates from ecological studies.

A number of commercial kits exist for the identification of yeasts. These are generally restricted either in the number of tests used or the range of yeasts covered. The small number of species associated with diseases of man has enabled reliable kits to be developed for their detection, and these are used in many public health laboratories. The cultural conditions in the kits differ from those in the tests on which current classification is based. It follows that results obtained with kits cannot reliably be used in conjunction with standard descriptions in taxonomic treatises to obtain accurate identifications. The kits may nevertheless be used to distinguish between strains where the pattern of behaviour under the conditions of the tests is known. A production strain, for example, may be exhaustively examined using a kit system so that the degree of reproducibility is well known. The kit may then be used routinely in quality control procedures, and any deviation from the expected test responses may indicate either an alteration in the perform- ance of the production strain or the introduction of contaminating microorganisms.

It cannot be stated too strongly that for a reliable identification, short cuts cannot be taken, and the test methods described for use with the different systems must be carefully followed. Specialist confirmation should be sought in cases where the accuracy of the identification is crucial.

5.2.6   Nomenclature

The morphological and physiological characteristics commonly used to define yeast taxa have been discussed, and Tables 5.1–5.3 list the genera and their properties. Species with known sexual or perfect states are termed teleomorphic whereas anamorphic or imperfect species lack this portion of their life cycle, at least in the laboratory. One of the cardinal points of taxonomy is that classification is based primarily on the teleomorphic state from which phylogeny can be inferred. If sexual reproduction is unknown, a species is classified in an anamorphic genus where members have similar morphology and physiology. For example, *Candida utilis* is the anamorph of *Pichia jadinii* while *Candida pulcherrima* has *Metschnikowia pulcherrima* as a teleomorph. Taxonomists recognise these two sexual genera to be phylogenetically distinct, but this difference is not readily apparent from their assignment to *Candida*. Consequently, anamorphic genera are a taxonomic convenience only to be used until a sexual state is found.

Synonomy presents a problem to those not familiar with classification. Synonyms result from changes in the nomenclature of valid species as well as from the inadvertent redescription of already known species. In the first instance, the food and feed yeast *Torula utilis* was reclassified as *Torulopsis utilis*. Later, it was transferred to *Candida* and became known as *Candida utilis*. Consequently, the names *Torula utilis* and *Torulopsis utilis* have become synonyms of *Candida utilis*. Now that *C. utilis* is recognised as the anamorph of *Pichia jadinii*, it becomes both the anamorph of this teleomorph as well as its synonym. *Candida guilliermondii* var. *nitratophila* is another synonym of *C. utilis*. This epithet, however, resulted from the naming of an unrecognised strain of *C. utilis* as the new variety of another species. Other examples include *Saccharomyces diastaticus* and *S. italicus* which were shown by DNA comparisons to be the same species as *S. cerevisiae*, and therefore become its synonyms. A knowledge of synonomy allows the biotechnologist to avoid some of the confusion resulting from poor classification. For example, discovery that a particular reaction can be mediated by *Saccharomyces cerevisiae*, *S. beticus*, *S. chevalieri*, *S. gaditensis*, *S. hispalensis* and *Zygosaccharomyces paradoxus* would suggest this to be a common metabolic function until we realise that all of these species are conspecific with *S. cerevisiae* and therefore represent synonyms.

The type specimen concept requires some mention because it is a vital part of the Code of Botanical Nomenclature, the internationally accepted rules governing the taxonomy of all plants and fungi. The Code requires that a dried specimen be deposited with a recognised institution as part of the validation process for describing new species. Because yeasts are fungi and therefore under the purview of the Code, the type material for a new yeast should be a dried herbarium specimen. Since yeast taxonomists generally require living cultures, they have adopted the practice of depositing living pure cultures termed type cultures or type strains. While this practice is technically illegal in terms of the Code and must ultimately be resolved, it does provide living reference material that one may use for taxonomic comparisons, and therefore forms the basis for present day yeast taxonomy.

5.3   Genetic and molecular relatedness among yeasts
5.3.1   *Genetic comparisons*
Early genetic studies, especially those by Winge & Roberts (1949) and Lindegren & Lindegren (1949), showed that a single gene could confer on a yeast the ability to ferment or assimilate a sugar, and this raised the possibility that species might be inadvertently and artificially established simply because of the functionality of one or a few genes. Following these findings was the discovery by Wickerham & Burton (1954) that ascospore shape in *Pichia ohmeri* might be either hat-shaped or spheroidal, depending on which complementary mating types were paired. While these studies cast some doubt on the validity of physiological tests and ascospore shape as means to separate species and genera, there were insufficient comparisons to justify discarding criteria that generally seemed to offer reliable resolution of species.

5.3.2   *Guanine+cytosine (G+C) content*
The first of the molecular criteria used to define species was the guanine+cytosine (G+C) content of the nuclear DNA. The taxonomic uses of G+C values are mainly exclusionary because the 600 or so known yeast species range in G+C content from approximately 28 to 70 mol%, and overlap between unrelated species is inevitable. The G+C values are usually determined from buoyant density in cesium chloride gradients generated by ultracentrifugation (Schildkraut, Marmur & Doty, 1962) or from thermal denaturation (Marmur & Doty, 1962). When determined by buoyant density, a difference in G+C content of 1.0–1.5 mol% or greater indicates strains to be different species (Price, Fuson & Phaff, 1978; Kurtzman *et al.*, 1980*a,c*), while this

difference is 2.0–2.5 mol% if determined from thermal denaturation (Meyer, Smith & Simione, 1978).

Examination of G+C values for a variety of taxa illustrates the interesting fact that the G+C contents of ascomycetous yeasts is about 28–50%, whereas that of basidiomycetous yeasts is approximately 50–70% (Nakase & Komagata, 1968; Kurtzman, Phaff & Meyer, 1983). Except for the narrow range of 48–52%, where some overlap occurs, the taxonomic class of imperfects can be reliably determined from their base composition. The range of G+C contents among species within a genus is quite often 10% or less as found in *Debaryomyces*, *Kluyveromyces* and many other genera. By contrast, species in *Pichia* differ by as much as 22%, and it seems likely for this and other reasons that the genus may be phylogenetically heterogeneous.

5.3.3   *Nuclear DNA relatedness*

Although G+C values have been helpful for separation of taxa, their uses are clearly limited. Considerably greater resolution is offered from comparisons of DNA complementarity as measured from reassociation reactions. The methodology has been described in detail elsewhere (Meyer & Phaff, 1972; Price, Fuson & Phaff, 1978; Kurtzman *et al.*, 1980c). The results from these comparisons are expressed as percent relatedness, but there has been some uncertainty in deciding the point at which strains are considered separate species. Several workers have suggested that relatedness above 70–80% demonstrates conspecificity (Martini & Phaff, 1973; Price, Fuson & Phaff, 1978).

Kurtzman and co-workers (1980a,c) examined this question through comparisons of heterothallic yeasts in which DNA complementarity can be fairly definitively correlated with fertility. In the first of these studies, *Pichia amylophila* and *P. mississippiensis* were shown to mate well, but ascospores were not viable. The two species showed only about 25% DNA relatedness, thus correlating lack of fertility with low DNA complementarity. However, the relationship between *Issatchenkia scutulata* var. *scutulata* and *I. scutulata* var. *exigua* was somewhat different. These two taxa also showed only about 25% DNA relatedness, but genetic crosses gave 3–6% viable ascospores. Furthermore, crosses between these F_1 progeny gave 17% viability among F_2 progeny. Reciprocal crosses among the F_1 progeny were fertile, as were backcrosses to the parents. From this, it appears that all stocks had essentially homologous chromosomes and that progeny were neither amphidiploids nor aneuploids. Clearly then, the lower limit of DNA–

DNA homology values suggesting species delimitation is not yet well defined, but base sequence divergence, as estimated from whole genome comparisons, may be as great as 75% before genetic exchange can no longer occur. Data in Table 5.7 summarises the preceding findings and provide further comparisons of other taxa. In these studies, progressively less DNA relatedness parallels decreasing mating competence and fertility. These data also indicate that resolution goes no further than the detection of sibling species. Beyond this, all species, regardless of extent of kinship, show essentially less than 10% complementarity.

However, despite the strong correlation between DNA relatedness and fertility exhibited in Table 5.7, exceptions may occur and whole genome DNA complementarity should therefore be regarded as a strong but not infallible indicator of biological relatedness. Exceptions to this trend would include chromosomal changes that affect fertility between strains such as inversions, translocations, autopolyploidy and aphidiploidy (allopolyploidy). Most of these changes would not be detectable in whole genome DNA comparisons. However, amphidiploidy is recognisable if parental species have been included in the study or should be suspected if genome sizes vary between strains showing substantial relatedness. For example, *Saccharomyces carlsbergensis* shows relatively high DNA homology with both *S. cerevisiae* and *S. bayanus* even though these latter two species demonstrate little relatedness (Fig. 5.6). A comparison of genome sizes suggests *S. carlsbergensis* to be a partial amphidiploid which arose as a natural hybrid of *S. cerevisiae* and *S. bayanus* (Vaughan Martini & Kurtzman, 1985). In this case, we would predict *S. carlsbergensis* to be infertile with the proposed parents despite relatively high DNA relatedness.

DNA relatedness studies have had an enormous impact on our assessment of the criteria used to define species and genera among the

Fig. 5.6. DNA relatedness between type strains of *Saccharomyces cerevisiae*, *S. bayanus* and *S. carlsbergensis* (after Vaughan Martini & Kurtzman, 1985). Relative genome size: *S. cerevisiae*, 1.00; *S. bayanus*, 1.15; *S. carlsbergensis*, 1.46.

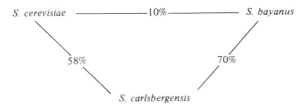

Table 5.7. *Correlation of mating reaction and DNA complementarity among closely related heterothallic ascomycetous and basidiomycetous yeasts*

Species	Mating reaction	% DNA relatedness
Filobasidiella neoformans × *Filobasidiella bacillispora*	fair conjugation; 0–30% basidiospore viability (F_1 progeny; F_2 not determined)	55–63[a]
Issatchenkia scutulata var. *scutulata* × *Issatchenkia scutulata* var. *exigua*	good conjugation; ascospores viable: F_1 = 5%; F_2 = 17%	21–26[b]
Pichia amylophila × *Pichia mississippiensis*	good conjugation; ascospores not viable	20–27[c]
Pichia bimundalis × *Pichia americana*	poor conjugation; ascospores not produced	21[d]
Pichia alni × *Pichia canadensis*	poor conjugation; ascospores not produced	6[e]
Issatchenkia orientalis × *Issatchenkia occidentalis*	infrequent conjugation; ascospores not produced	3–8[b]

[a] Aulakh, Straus & Kwon-Chung (1981).
[b] Kurtzman et al. (1980a).
[c] Kurtzman et al. (1980c).
[d] Kurtzman (1984a).
[e] Phaff, Miller & Miranda (1979); Fuson et al. (1979).

yeasts (Table 5.8). Yarrow & Meyer (1978) combined the genera *Candida* and *Torulopsis* when it was found that presence or absence of pseudo-hyphae might occur in a single species. Originally, these anamorphic genera were separated on the basis that species of *Candida* produced pseudohyphae whereas species of *Torulopsis* did not. The genera are now regarded as synonymous and *Candida* has priority of use because it was described first. Similarly, the 75% relatedness detected between *Pichia lindneri* and *Hansenula minuta* prompted Kurtzman (1984*b*) to propose that the two genera, which are separated on ability to assimilate nitrate, be combined. The DNA comparisons have also complemented earlier genetic studies in their demonstration that differences in fermentation and carbon assimilation are common among strains of a species.

5.3.4 *Mitochondrial DNA relatedness*

Comparisons of mitochondrial DNA (mtDNA) provide another means for examining taxonomic relationships. Relatedness has been examined through comparisons of fragment patterns generated with restriction endonucleases. The small size of the mitochondrial genome makes such comparisons practical, whereas the larger nuclear genome

Table 5.8. *DNA relatedness between yeast species differing in traditional taxonomic characteristics*

Species	Characteristic/+ or −	% DNA relatedness
Candida slooffii	pseudohyphae/+	80[a]
Torulopsis pintolopesii	/−	
Hansenula wingei	true hyphae/+	78[b]
Hansenula canadensis	/−	
Debaryomyces formicarius	glucose ferm/+	96[c]
Debaryomyces vanriji	/−	
Schwanniomyces castellii	lactose assim/+	97[c]
Schwanniomyces occidentalis	/−	
Hansenula minuta	nitrate assim/+	75[d]
Pichia lindneri	/−	
Sterigmatomyces halophilus	nitrate assim/+	100[e]
Sterigmatomyces indicus	/−	

[a] Mendonça–Hagler & Phaff (1975).
[b] Fuson *et al.* (1979).
[c] Price *et al.* (1978).
[d] Kurtzman (1984*b*).
[e] Kurtzman *et al.* (1980*b*).

would present more fragments than could be reasonably handled. McArthur & Clark-Walker (1983) used mtDNA restriction patterns to correlate perfect–imperfect relationships between the yeast genera *Dekkera* and *Brettanomyces*. They found identical restriction patterns for the pair *D. bruxellensis/B. lambicus*, as well as for the imperfects *B. abstinens/ B. custersii* and *B. anomalus/B. clausenii*. This strongly suggests the pairs to be conspecific. Size differences in mtDNA among the other species assigned to these genera prevented an unambiguous assessment of their relationships. However, it appears that the taxonomic resolution from mtDNA restriction patterns is no greater than afforded by whole nuclear DNA reassociations.

5.3.5   *Ribosomal RNA relatedness*

The rather narrow resolution provided by whole genome DNA reassociation and restriction analysis does not allow verification of species assignments within genera or an understanding of intergeneric relationships. The DNA coding for ribosomal RNA (rRNA) appears to be among the most highly conserved sequences known, and it offers a means for assessing affinities above the species level. One of the commonest methods for measuring rRNA complementarity consists of immobilising single-stranded DNA onto nitrocellulose filters and incubating the filters in a buffer solution with radiolabelled rRNA (Bicknell & Douglas, 1970; Kennell, 1971; Johnson, 1981). A modification of this method was developed by Baharaeen, Melcher & Vishniac (1983) in which complementary DNA was synthesised on 25S rRNA fragments and then allowed to hybridise with rRNA in solution. As we will discuss, another method that holds promise for phylogenetic studies is the comparison of sequences from rRNA subunits.

Woese and collaborators (Fox *et al.*, 1980) have effectively assessed phylogenetic relationships among the prokaryotes through cataloguing 16S rRNA oligonucleotides generated with T$_1$ ribonuclease. Johnson & Harich (1983) reported good correlation' of values obtained by this method with those from experiments in which rRNAs are hybridised to DNA bound on nitrocellulose. Application of the oligonucleotide cataloguing method to fungi is unlikely in the near future because of the greater number of fragments generated from the 18S ribosomal subunit of eukaryotes (G. E. Fox, personal communication).

Bicknell & Douglas (1970) were among the first to employ rRNA comparisons to assess phylogenetic relatedness among fungi. Their study focused on the genus *Saccharomyces*, some species of which now have been assigned to the genera *Zygosaccharomyces, Kluyveromyces* and *Torulaspora*. In general, the more highly related species clustered into groups corresponding to present generic assignments. Similar results were obtained for this same group of species by Adoutte-Panvier *et al.* (1980) through electrophoretic and immunochemical comparisons of ribosomal proteins. Bicknell & Douglas (1970) also showed that 25S rRNA comparisons cannot resolve the phylogeny of closely related species because of the highly conserved nature of the sequences.

Walker & Doolittle (1982, 1983) used 5S rRNA sequences to investigate relatedness among the basidiomycetes, including several yeast taxa. Their work identified two distinct clusters that correlated with the presence or absence of septal dolipores rather than with the traditional separation of these species into the classes Heterobasidiomycetae and Homobasidiomycetae. Further, the data suggest that capped dolipores evolved from capless dolipores which may have evolved from single septal pores. In continuing work, Walker's (1985) analysis of 5S rRNA sequences from several ascomycete genera suggests much greater divergence than previously suspected. Blanz & Gottschalk (1986) have provided comparisons of 5S rRNA from certain of the lower basidiomycetes.

5.3.6 *Nucleic acid relatedness compared with other methodologies*

Examples have been given in which species defined by traditional methods did not coincide with those determined from comparisons of nucleic acids. Taxonomic conclusions drawn from other methodologies, such as numerical analysis or from proton magnetic resonance (PMR) spectra and serology of cell wall mannans, have sometimes been at odds with data from nucleic acids studies. For example, using numerical analysis, Campbell (1973) proposed reduction of *Hansenula wingei, H. canadensis*, and both varieties of *H. bimundalis* to a single species. Fuson and co-workers (1979) showed from DNA sequence studies that *H. canadensis* and *H. wingei* are conspecific, whereas this latter species and *H. bimundalis* show little relatedness. Kurtzman (1984*a*) demonstrated only 21% complementarity between the two varieties of *H. bimundalis*. This disparity of the two methods probably results from the relatively limited amount of phenotypic data available for numerical analysis, whereas DNA studies have the whole genome as a database. This also appears to be the case for separations made using proton magnetic resonance, since Spencer & Gorin (1969) found the spectra of cell wall mannans from *H. beckii* and *H. canadensis* to be quite similar, but different from the pattern shared by *H. wingei* and the varieties of *H. bimundalis*. In further comparisons of this group, Tsuchiya *et al.* (1974) reported the cell surface antigens of *H. wingei*, *H. canadensis*, and *H. beckii* to be indistinguishable. The reason for this discrepancy is not at all clear, but Ballou (1974) has shown for *Saccharomyces* mannans that single gene changes can impart significant differences in immunological reaction. The difference between the two main PMR spectral types of mannan in *S. cerevisiae* was also found to be controlled by a single gene (Spencer, Gorin & Rank, 1971).

Differentiation of many ascosporogenous genera is based on ascospore shape and surface structure. An opportunity to test the soundness of this convention in the genera *Schwanniomyces*, *Saccharomyces*, *Debaryomyces* and *Pichia* arose through a comparison of the DNA data of Price, Fuson & Phaff (1978) with the scanning electron microscopy studies by Kurtzman, Smiley & Baker (1972, 1975), Kurtzman & Smiley (1974, 1979) and Kurtzman & Kreger-van Rij (1976). All four species of *Schwanniomyces* had high DNA relatedness (Price *et al.*, 1978) and similar spore architecture (Kurtzman, Smiley & Baker 1972; Kreger-van Rij, 1977). The least related, *S. persoonii* at 80% complementarity, showed fewer and less-pronounced spore protuberances than the other three species, all four of which are now regarded as conspecific. Ascospore topography for many species of *Torulaspora* was essentially identical and also frequently indistinguishable from the majority of species in *Debaryomyces*. Although many species within their respective genera were found to be conspecific by DNA homology, others proved distinct despite spore similarity. The few species with recognisably different spores showed little DNA relatedness. Thus, species with unlike spores can be expected to show little DNA complementarity, but no prediction can be made concerning species with similar spores. A few exceptions exist among other groups. Ascospores of *Pichia ohmeri* may be either spheroidal or hat-shaped, depending on mating type (Wickerham & Burton, 1954; Fuson, Price & Phaff, 1980), and spore shape in *Yarrowia lipolytica* is mating-type-dependent (Wickerham, Kurtzman & Herman, 1969). Spore shapes among *Kluyveromyces* spp. with high degrees of DNA complementarity may be either spheroidal or kidney-shaped (H. L. Presley & H. J. Phaff, personal communication).

Kreger-van Rij & Veenhuis (1973) demonstrated by transmission electron microscopy a basic difference in hyphal septa among ascomycetous yeasts. Three classes of septum ultrastructure were observed: a single central pore (micropore), plasmodesmata and a dolipore configuration. All mycelial species of *Hansenula* and *Pichia* have the micropore, whereas plasmodesmata are common to all *Saccharomycopsis* species except *S. lipolytica* (=*Yarrowia lipolytica*). Dolipores were found in species of *Ambrosiozyma*. Future comparisons of these species by molecular techniques that detect distant relatedness might show ultrastructural differences in septa to be of evolutionary significance, as now appears to be the case among the lower basidiomycetes (Walker & Doolittle, 1982, 1983).

One issue that needs attention concerns the means for detecting phylogenetic relationships of species within genera. As we have seen, whole-genome DNA complementarity is too specific, whereas rRNA comparisons may be too broad in resolution to detect closely related species (Bicknell & Douglas, 1970). However, the DNA coding for certain enzymes such as glutamine synthetase and superoxide dismutase, although less conserved than that coding for rRNA, is still less divergent than is apparent from comparisons of whole-genome DNA (Baumann & Baumann, 1978; Baumann, Bang & Baumann, 1980). By determining changes in the amino acid sequence of these enzymes through such techniques as quantitative microcomplement fixation, an estimate of intermediate relatedness may be found. Immunological studies of protein similarity are very sparse for yeasts, and the resultant information is highly dependent on the kinds of proteins studied. Lachance & Phaff (1979) used exo-β-glucanases from species of *Kluyveromyces* but found that the enzyme was poorly conserved and that immunological distances between most species were too great to be reliable for determining evolutionary relationships in the genus.

Electrophoresis of allozymes presents another means for estimating molecular diversity and, as with immunological studies, resolution depends on the extent of sequence conservation. The proportion of point mutations that are electrophoretically detectable is estimated at approximately 0.27, because of the redundancy in the genetic code and the large proportion of amino acids that are electrically neutral (Baptist, Shaw & Mandel, 1971; Selander, 1976; Holzschu, 1981). The nearly universal occurrence of extensive protein polymorphism in natural populations has led some workers to believe that variation would prove physiologically important and, therefore, under selective control; others regard it as without phenotypic effect and thus selectively neutral (Selander, 1976). Regardless, data derived by this technique do provide insight into evolutionary processes and taxonomy. Baptist & Kurtzman (1976) utilised comparative enzyme patterns to separate sexually active strains of *Cryptococcus laurentii* var. *laurentii* from non-reactive strains and from the varieties *magnus* and *flavescens*. Yamazaki & Komagata (1981), utilising similar techniques, examined species relationships in *Rhodotorula* and its teleomorph *Rhodosporidium*. Holzschu (1981) studied the evolutionary relationships among some 400 strains of various cactophilic species of *Pichia*. His study of the banding patterns in starch gels of 14 metabolic enzymes allowed a determination of the genetic distances among the various yeast populations.

These examples show that yeast taxonomy is still in a state of flux but that many problems have been resolved through molecular comparisons. These new findings result in name changes that are sometimes inconvenient and that may even seem capricious, but that are necessary if our taxonomy is to reflect phylogeny, a supposition implicit in the assignment of species to genera and of genera to higher orders of classification.

5.4   Yeasts and yeast-like fungi of industrial significance and their taxonomy

5.4.1   *Introduction*

The majority of yeasts presently of industrial use are teleomorphic and anamorphic forms of ascomycetes. For example, the primary species involved in baking and brewing is *Saccharomyces cerevisiae*, but *S. bayanus* and *S. carlsbergensis* are other closely related species sometimes used. While basidiomycetous yeasts seldom produce a gaseous fermentation of sugars, their capabilities for metabolism of a wide variety of compounds suggest that they represent a basically untapped resource for future industrial exploitation. Examples of industrial uses of yeasts are given in this section and the characteristics of the taxa can be reviewed in Tables 5.1–5.3.

5.4.2   *Food and feed yeasts*

The use of yeasts for production of bread products and alcoholic beverages does not represent their only food use. *Saccharomyces cerevisiae* from alcohol fermentations is added to foods and animal feeds as a source of protein and vitamins. The main impediment to using large amounts of yeast in the diet is the high RNA content of the cells which can lead to clinical manifestations such as gout. Cell wall mannans are added to foods to improve texture (Robbins & Seeley, 1978), and the mannans can also be used as carriers of flavouring agents.

Oriental food fermentations frequently rely on a variety of microorganisms. The soy sauce fermentation requires certain moulds and a bacterium as well as *Zygosaccharomyces rouxii* and species of *Candida* (synonym *Torulopsis*) (Hesseltine, 1983). The so-called Chinese yeast, an amylolytic starter of Lao chao, generally contains large numbers of the starch hydrolysing species *Saccharomycopsis fibuligera* (Hesseltine, 1983).

Candida utilis (teleomorph=*Pichia jadinii* (Kurtzman, Johnson & Smiley, 1979)) is another commonly used food and feed yeast. This species can assimilate pentoses, and its tolerance to lignin degradation products makes it ideal for growth in paper pulping wastes. In fact, *C. utilis* has a long history as a food yeast and was grown on wood-derived xylose in Germany during World Wars I and II (Lindner, 1922, as cited by Pyke, 1958; Peppler, 1970).

An additional use of *Candida utilis* has been in conjunction with *Saccharomycopsis fibuligera* to dispose of starchy wastes. The basic procedure was developed by Wickerham *et al.* (1944) and was put into practice in Sweden as the Symba process to dispose of potato processing wastes (Jarl, 1969). *Saccharomycopsis fibuligera* can assimilate starch as a carbon source, but its growth rate is slow. By contrast, *C. utilis* has a rapid growth rate but cannot assimilate starch. When grown together, cross-feeding of *C. utilis* results in a substantial amount of yeast cells, which are used as an animal feed supplement.

Cheese whey represents another waste that may be fermented. Disposal frequently has been through municipal sewage systems or by spraying onto fields. The high lactose content of whey has made it attractive for the production of ethanol in dairying areas. The species most commonly used for fermentation is the yeast *Kluyveromyces marxianus* (synonym, *K. fragilis*). *Saccharomyces cerevisiae*, commonly used in many other fermentations, is unable to ferment lactose. Before fermentation, the whey or whey permeate must be supplemented with nitrogen, usually in the form of ammonia, as well as with vitamins and minerals. Alcohol yields do not exceed 3–6% (Bothast *et al.*, 1986), making this a marginal economic process if the disposal cost of untreated whey is not taken into account. Yeast cells are recovered from the fermentation and may be used in food or animal feed. Annual production of yeast cells from this process is estimated to be around 5000 tons.

5.4.3   *Enzymes*

As we discussed in the previous section, *Saccharomycopsis fibuligera* is amylolytic and can serve as a commercial source of amylase. A number of other species also produce amylase and these can be found by looking for starch assimilation among the physiological reactions given for each species listed in *The Yeasts, A Taxonomic Study* (Kreger-van Rij, 1984). This book also gives sources of strains, and such information generally allows selection of non-pathogenic isolates with the desired properties.

Other enzymes of commercial importance include maltase from *Saccharomyces italicus* (Cooney & Schaefer, 1982), a synonym of *S. cerevisiae* (Vaughan Martini & Kurtzman, 1985), and lactase from *Kluyveromyces fragilis* (=*K. marxianus*) and *K. lactis* (Fenton, 1982). One use of the purified enzyme is the hydrolysis of lactose in milk products for people with lactose intolerance. Taxa such as *Candida lipolytica* (teleomorph=*Yarrowia lipolytica*) and species of *Trichosporon* produce a variety of proteases. These same species are also noted for significant amounts of lipase (Wickerham *et al.*, 1969; Terada, 1972). Other commercially important enzymes include acyl-coenzyme A oxidase from *Candida* sp. (Kikuchi, Ogawa & Ando, 1983) and endo-deoxyribonuclease A from a species of *Saccharomyces* (Ando, Shibata & Watabe, 1984). Fukumura (1976) showed *Cryptococcus laurentii* to give a nearly 100% conversion of L-aminolactam into L-lysine, a process of considerable economic potential.

5.4.4   *Production of various biochemicals*

A wide variety of biochemicals are synthesised by yeasts although ethanol is the major compound produced commercially. In the US alone, 2,650 million litres of ethanol are distilled annually for supplementation of motor fuels (Murtagh, 1986). *Saccharomyces cerevisiae* is the species most commonly used to ferment sucrose and its component hexoses, but pentoses from plant residues may be fermented by *Pachysolen tannophilus* (Schneider *et al.*, 1981; Slininger *et al.*, 1982), *Pichia stipitis* (Toivola *et al.*, 1984) and *Candida shehatae* (du Preez & van der Walt, 1983). The main disadvantage with pentose fermentations has been low ethanol yields, about one-third that obtained from hexoses fermented by *Saccharomyces*.

The discovery in the late 1960s that high yields of citric acid were produced by *Candida lipolytica* when grown on *n*-paraffins (Yamada, 1977) signalled a potential dramatic change in the fermentation industry because this organic acid has been commercially produced since 1919 by *Aspergillus niger*. Yields generally exceeded 150 g/l, or about twice that synthesised by *A. niger* on sucrose. This finding occurred at a time when sugar substrates were being increasingly diverted to food use and petroleum was still inexpensive. Whether *C. lipolytica* replaces *A. niger* depends upon the price of petroleum and the concern that petroleum-derived carcinogens might be present in the final product. Other organic acids, such as isocitric and pyruvic, are produced in large quantities by *C. lipolytica* making this quite an industrially important yeast (Yamada, 1977).

Straight chain hydroxy fatty acids have been isolated in moderate quantities from the fermentation broth of several yeasts. Both *Saccharomycopsis fibuligera* and *Pichia sydowiorum* (synonym *Hansenula sydowiorum*) give mixtures of C_{14}–C_{18} 2-D-hydroxy acids (Vesonder *et al.*, 1970; Kurtzman, Vesonder & Smiley, 1973) while *Saccharomycopsis malanga* produces 3-D-hydroxypalmitic acid (Kurtzman, Vesonder & Smiley, 1974). These metabolites occur as crystals in the culture media which allowed their initial detection by observation under the light microscope. Brodelius, Nilsson & Mosbach (1981) reported the formation of α-keto acids by *Trigonopsis variabilis*. These particular compounds are of use in the therapy of patients suffering from acute uremia because excess nitrogen is removed from the body through transamination of the keto acids.

A variety of other compounds from yeasts are produced in sufficient quantity to be of industrial significance. Among these are D-arabitol from *Pichia haplophila* (Fujiwara & Masuda, 1981) and D-mannitol from *Candida lipolytica* grown on hydrocarbons (De Zeeuw & Tynan, 1973). *Pichia ciferrii* (synonym *Hansenula ciferrii*) produces extracellular crystalline tetraacetylphytosphingosine (Stodola & Wickerham, 1960; Wickerham & Stodola, 1960; Wickerham & Burton, 1962). The sphingosines are also found in higher plants and the nerve tissue of animals.

Extracellular microbial polysaccharides have served as important industrial replacements of plant gums. Although yeast polysaccharides have not yet found much commercial use, species with heavy capsules occur among both the ascomycetes and the basidiomycetes. Phaff (1971) and Slodki (1980) summarised known structures and physical properties of polysaccharides from a large number of yeasts. Such data allow some prediction of potential applications.

5.4.5 *Yeast-like fungi of industrial use*
The distinction between yeasts and other fungi was discussed earlier, but among this latter group are species termed yeast-like. Such species are common and may be dimorphic higher taxa that form yeast-

like growth through budding, fission or successive conidiation, or they may be species that have simple, unenclosed fruiting bodies which, for other reasons, are not considered as part of the yeasts. *Aureobasidium pullulans* and many other dematiaceous hyphomycetes are among those representing the former group, while *Ashbya, Eremothecium, Ascoidea, Dipodascus, Cephaloascus, Taphrina, Tilletia, Ustilago* and similar genera form unenclosed teleomorphs and represent the latter group.

Certain of these yeast-like species find considerable industrial use. The vitamin riboflavin is produced almost exclusively by fermentation with *Ashbya gossypii* (Wickerham, Flickinger & Johnston, 1946; Demain, 1981). *Eremothecium ashbyi* is another strong riboflavin producer. *Geotrichum candidum* (teleomorph=*Endomyces geotrichum*), common to soil and some dairy products, is used for commercial production of lipase. Of the dimorphic higher fungi, *Aureobasidium pullulans* produces the extracellular polysaccharide pullulan which can be utilised as a plastic-like, biodegradable packaging material (Duffy, 1980). Colour variants of *A. pullulans* from tropical and subtropical areas are noted for their production of xylanase, an enzyme of great interest for conversion of biomass to chemical feedstocks (Leathers, Kurtzman & Detroy, 1984).

6

Patent protection for biotechnological inventions

I. J. BOUSFIELD

6.1   Introduction

This chapter is intended to give the reader who is unfamiliar with patents an introduction to the patent system as it applies to biotechnology, and a general guide to the procedures and pitfalls involved in obtaining patent protection for biotechnological inventions. For a detailed discussion of the whole subject of patents in biotechnology and a review of the variety of national patent systems the reader is referred to the excellent texts by Crespi (1982), Beier, Crespi & Straus (1985) and Straus (1985). It is not possible here to provide a step-by-step guide to getting a patent in every country in the world, for, despite an overall similarity, variations between different national patent laws are manifold, and professional help is necessary to guide even the experienced inventor through their complexities. The present account does no more than skim the surface of what is a complex and often fascinating subject; for this reason a short list of selected publications which illustrate in more detail many of the points raised here is given in Section 6.6, Further reading.

6.2   Basis of the patent system
6.2.1   *Principles*

The principle (if not the practice) of the patent system is straightforward: the inventor of a new product or process publicly discloses the details of his invention and in return he is granted for a limited period a legally enforceable right to exclude others from exploiting it. In this way the inventor's ingenuity is acknowledged and rewarded, while at the same time further technical progress is encouraged by the public dissemination of information about the invention.

6.2.2   *Criteria for patentability*

To qualify for patent protection, an invention must meet the following major criteria.

Novelty. The invention must be *new*. Most countries apply the test of 'absolute novelty', which means that if prior knowledge of it exists anywhere in the world (not merely in the country where patent protection is sought) then the invention belongs to the state of the art ('prior art') and is not patentable. The prior art is held in these countries to include anything the inventor himself may have said or written about his invention. Exceptions to this rule of absolute novelty are the patent systems of the USA and Canada, where publications by the inventor made not more than one (USA) or two (Canada) years before a patent application is filed in that country do not destroy novelty. 'Grace periods' of six months are also allowed by Australia, New Zealand and Japan, but only in respect of certain kinds of publications made by the inventor, for example at certain scientific meetings. Under nearly all patent systems, publications made after the date that the patent application is filed (the 'priority date') do not jeopardise protection for that particular subject matter in that particular application. It must be remembered, however, that they will form part of the prior art against which any *future* applications will be assessed.

Inventiveness. The invention must show evidence of an *inventive step*, that is it must not be 'obvious' from the state of the art to anyone 'skilled in the art'. In simple terms this means that the average expert in the field under consideration could not reasonably have predicted the invention as an obvious or logical outcome of what he already knew.

Utility (industrial applicability). The invention must have a *practical use*, which in the USA means exactly that, but in nearly all other patent systems means that it must also be capable of industrial application. Most countries, however, hold that medical methods for the direct treatment of the human or animal body are not susceptible to industrial application and are therefore not patentable. The major exception to this is again the USA.

Disclosure. The details of the invention must be *disclosed* by means of a patent specification (see Section 6.5.2 below) so that the invention is described in sufficient detail to allow a skilled person to reproduce it.

This is normally done by means of a written description supplemented where necessary by drawings. However, in one major category of biotechnological inventions – those involving the use of new microorganisms and certain other novel living materials – a written description is not usually considered sufficient for the purposes of disclosure. In such cases it is argued that no matter how carefully the description may be worded, if the microorganism itself is not available, the invention cannot be reproduced. Therefore, many countries require the deposit of new microorganisms in a recognised culture collection to ensure their public availability. This unique aspect of biotechnological patent procedure is dealt with in some detail later in this chapter.

Exclusions from patentability. As well as meeting the criteria listed above, an invention must not be of a kind which is excluded from patentability by its very nature. Exclusions relating specifically to biotechnological inventions are discussed later, but in general terms patents cannot be obtained for mere discoveries, theories, computer programs, literary works, musical compositions, aesthetic creations and illegal or offensive devices.

6.3   Kinds of biotechnological invention

There are four main kinds of biotechnological invention: products, compositions, processes, and use or methods of use (Crespi, 1982).

6.3.1   *Products*

These inventions are exactly what the word suggests – tangible new materials or entities. They include organisms themselves (e.g. bacteria, fungi), parts of organisms (e.g. cell lines), substances produced by either of these (e.g. enzymes, antibiotics), and substances obtained by or employed in recombinant DNA techniques (e.g. plasmids, DNA molecules).

Product inventions can be the subject of two broad kinds of patent claim: the 'product *per se*', where patent protection is sought for the product itself, regardless of the method of manufacture, and the more limited 'product-by-process', where protection is sought for the product obtained by a particular process.

6.3.2   *Compositions*

These inventions are mixtures of substances or organisms, the individual components of which may already be known, but which in combination can be shown to display a new property or exert a new effect.

6.3.3   *Processes*

These inventions are methods for the manufacture of products, and include bioconversions, fermentations, and methods of isolation, purification or cultivation. Some process inventions are genuinely new methods for obtaining novel or known products, but others are known methods applied in new situations or used in the manufacture of novel products.

6.3.4   *Use and methods of use*

Methods of use include processing or treating materials (e.g. industrial raw materials or agricultural products), non-medical treatments of humans or animals, 'off the body' medical methods (e.g. a method of diagnosis carried out on a sample taken from a patient), methods of testing (e.g. quality control) and in a few countries, notably the USA, medical treatment of humans and animals. Also, the new medical use of a substance previously unknown to have that use is protectable in European Patent Convention (EPC) countries.

6.4   Patentability of biotechnological inventions

6.4.1   *Inventions involving new plants and animals*

Plant varieties. By far the most common form of legal protection for new plant varieties is the plant variety right (although in the USA special 'plant patents' are available for asexually propagated plants). Several countries are now party to the International Convention for the Protection of New Varieties of Plants (UPOV) which aims to harmonise national practices as far as possible (International Convention, 1978). In these countries, plants that are protected by plant variety rights are usually specifically excluded from patentability. Plant variety rights in general are intended to allow the commercial plant breeder a monopoly on the production of propagating material for the purposes of commercial marketing, its offer for sale and its marketing. Plant variety rights are easier to obtain than is patent protection as there is no requirement for inventiveness or for reproducible disclosure, but they are more limited in scope in that neither the plant itself nor consumables produced from it (e.g. fruit for eating, grain for milling) are protected.

Plant variety rights were introduced essentially to cover varieties developed by traditional breeding methods, and it is such varieties that are excluded from patentability in many countries. Thus Article 53(b) of the EPC, to which 13 countries belong (Table 6.1) states the following:

> European patents shall not be granted in respect of . . . plant or animal varieties or essentially biological processes for the pro-duction of plants or animals; this provision does not apply to microbiological processes or the products thereof.

The same exclusion is found where the national laws of individual European countries have been harmonised with the EPC and is also contained in the laws of several other countries, e.g. the German Democratic Republic, Mexico, Sri Lanka, Thailand and Yugoslavia. Plant and animal varieties, but not essentially biological processes, are also excluded from patentability in China.

The exclusion of plant varieties from patent protection is a contentious issue. Straus (1985) has pointed out that current systems for the protec-tion of plant varieties were introduced when plant breeding methods did not permit the breeder to fulfil the normal criteria of patentability. However, the advent of new technologies, particularly genetic manipulation techniques, for the production of new plant varieties has

Table 6.1. *Countries party to the European Patent Convention at 1 January 1987*

Austria
Belgium
France
Germany (Federal Republic)
Greece
Italy
Liechtenstein
Luxembourg
Netherlands
Spain
Sweden
Switzerland
United Kingdom

meant that these requirements can now be met. Therefore, there seems to be a good argument in favour of allowing the developer of such varieties the right to choose between protection under the patent system or through plant variety rights (Beier *et al.*, 1985; Straus, 1985).

Animal varieties. Although a special form of protection for new animal varieties (which are not, however, regarded as inventions) is available in the Soviet Union, there is in general no special system of legislation for their protection. Most of the countries that exclude plant varieties from patentability also exclude animal varieties, and in those countries that do not, the position is not altogether clear. However, recent court decisions in the USA (*in re* Diamond & Chakrabarty, 1980) and Canada (*in re* Abitibi, 1982) suggest that animals may be patentable provided that the requirements for enabling disclosure, that is repeatability, are met (Straus 1985).

Processes for the production of animals and plants. Many patent laws, including those of the EPC, specifically deny patent protection for 'essentially biological processes' for the production of plants or animals; microbiological processes, however, are not included in this provision. This terminology may not be entirely clear and perhaps needs some explanation. Stated simply, an essentially biological process is considered to be one in which the result is achieved with a minimum of human technical intervention. The example given by the European Patent Office (EPO) in its guidelines to examiners is a method for selectively breeding horses in which human intervention is limited to bringing together animals having particular characteristics. On the other hand, a process for treating a plant to promote or suppress its growth, e.g. a method of pruning or of applying stimulatory or inhibitory substances, would not be considered to be essentially biological, since although a biological process is involved, the essence of the invention is technical.

Given this definition of 'essentially biological' and the exemption of microbiological processes from it, the normal criteria for patentability can be applied to methods for producing plants by, for instance, genetic manipulation involving the use of vectors in microbial hosts, or by somatic cell hybridisation. However, matters are less certain under the EPC in respect of some processes for the production of new animal varieties, even though such processes may meet the test of not being essentially biological. This is because a further exclusion from patentability is found in Article 52(4) of the EPC, which states the following:

> Methods for treatment of the human or animal body by surgery or therapy and diagnostic methods practised on the human or animal body shall not be regarded as inventions which are susceptible of industrial application . . . [See also Utility in Section 6.2.2 above.]

Straus (1985) has expressed the fear that some present and future approaches to animal breeding, such as techniques of embryo transfer, could be denied patent protection by this provision. In support of his argument he cites a recent decision by the UK Comptroller of Patents, in which an application involving just such a technique was rejected as being a method of treatment by surgery.

In contrast to the European system, the patent laws of the USA, Japan and China do not exclude essentially biological processes. Furthermore, the US laws do not exclude methods for treating animals (or humans); therefore the problems presented by the European system in respect of patenting processes involved in animal breeding do not exist in the USA.

Tissue cultures. Animal cell lines and plant tissue cultures (and in Japan, seeds) are generally considered to be in the same category as microorganisms for patent purposes. Thus they are subject to the provisions applied to microbiological inventions as discussed below. As regards plant cells, however, the US Patent Office makes a distinction between undifferentiated cell lines used, for instance, to produce a particular substance, and cells which are capable of differentiation and which are used simply to reproduce the whole plant.

6.4.2 *Inventions involving microorganisms*

Applied microbiology in its broadest sense is a major facet of modern biotechnology, and any discussion of patents in biotechnology inevitably must focus on the peculiar problems posed by microbiological inventions. In fact so great has been the attention given to these problems in patent circles that the patent legislation of an increasing number of countries contains specific provisions for inventions involving the use of microorganisms, and one international convention (the Budapest Treaty; see below) deals entirely with microorganisms (Budapest Treaty and Regulations, 1981).

It should be said that the term 'microorganism' is used in patent circles in a very loose sense and encompasses entities that strictly speaking are not microorganisms, e.g. cell lines and plasmids. Indeed, the word is intentionally not defined in the Budapest Treaty so as to avoid undue constraints being imposed upon the application of the Treaty, and in the words of the World Intellectual Property Organisation (WIPO) commentary on the draft Treaty (WIPO, 1980), it 'need not correspond to usage in some scientific circles'. Unfortunately, the use of such inexact terminology has led to uncertainty in some quarters as to what is or is not a microorganism. Because of this, the present author (acting on behalf of the World Federation for Culture Collections (WFCC) Patents Committee) has proposed to WIPO that the expression 'living material' be used instead of the word microorganism, particularly in regard to the Budapest Treaty. The word 'living' was defined as 'that material which under appropriate conditions is able to replicate itself, or which at least possesses the functional genes necessary to direct its own replication'. This definition has two advantages: first, it avoids insoluble philosophical arguments about where chemical reactivity ends and life begins, and second, it excludes non-living biological materials such as enzymes. In the present chapter, although the word microorganism is used for ease of reference, it should be taken to mean living material as defined above.

Microbiological inventions may be found in all of the categories of biotechnological invention outlined in Section 6.3 above. In general, microbiological processes and the (inanimate) products obtained by them can be considered analogous to chemical processes and products, and obtaining patent protection for them in most countries is nowadays fairly straightforward, provided that the basic criteria for patentability are fulfilled. Less straightforward, however, is the patenting of microorganisms as products either *per se* or as products-by-process. It is these inventions above all others that demonstrate the difficulties of determining the borderline between 'discoveries' and 'inventions', of what is 'new' and what is not, and of ensuring sufficiency of disclosure.

Naturally occurring microorganisms. A previously unknown naturally occurring microorganism that is left in its natural state is universally regarded as a discovery and is unpatentable. However, the degree of human intervention considered necessary to turn such a discovery into a patentable invention (assuming it has a practical use) varies between different countries. The extent of this variation was demonstrated by the official replies to a questionnaire on patent protection in biotechnology distributed to governments in 1982 by the Organization for Economic Cooperation and Development (OECD). These responses were reviewed in detail by Crespi (1985). In those countries that do permit naturally occurring organisms to be patented, isolation and purification of the organism are general prerequisites, after which various constraints are applied, mainly relating to novelty and unexpected properties. Thus, for example, the UK and the EPO require an organism to be 'new' in the sense of being hitherto *unknown*, whereas in Canada a new organism is one that does not already *exist* in nature (in this connection, Crespi (1985) has commented pointedly on the illogicality of equating 'unknown' with 'not previously existing'); the Federal Republic of Germany requires that 'certain changes occur during isolation so that the isolated microorganism is not identical with that occurring in nature'; Denmark requires naturally occurring organisms to have unforeseen properties. The USA permits the patenting of naturally occurring organisms as 'biologically pure cultures'.

Non-naturally occurring microorganisms. After the much-publicised Chakrabarty case in the USA in 1980, in which a genetically manipulated strain of *Pseudomonas* was held not to be a product of nature but a human invention patentable *per se*, there are unlikely to be any unusual problems in obtaining patent protection for 'artificial' microorganisms (bacterial recombinants, hybridomas, etc.), other than in countries which do not permit the patenting of any kinds of microorganism.

Sufficiency of disclosure. As already mentioned (Disclosure, in Section 6.2.2 above), one of the fundamental requirements of the patent system is that the details of an invention must be disclosed in a manner sufficient to allow a skilled person to reproduce the invention. Microbiological inventions present particular problems of disclosure in that more often than not repeatability cannot be ensured by means of a written description alone. In the case of an organism isolated from soil, for instance, and perhaps 'improved' by mutation and further selection, it would be virtually impossible to describe the strain and its selection sufficiently to guarantee another person obtaining the same strain from soil himself. In such a case, the strain itself forms an essential part of the disclosure. In view of this an increasing number of countries require a 'new' microorganism (i.e. one not already generally available to the public) to be deposited in a recognised culture collection whence it can subsequently be made available at some stage in the patent procedure.

As a general principle, an invention should be reproducible from its description at the time that the patent application is filed. In the case of an invention involving a new organism, therefore, most patent offices require a culture of the organism to be deposited not later than the filing date of the application (or the priority date if priority is claimed from an earlier application – see Section 6.5.3 below). Exactly when a strain becomes available varies according to the patent laws of different countries, and is a much debated question dealt with in more detail below (Release of samples). Since an invention must also be reproducible throughout the life of the patent, a microorganism deposited for patent purposes must remain available for at least this length of time. Most countries provide for a considerable safety margin in this respect, and availability for at least 30 years is a common requirement.

The Budapest Treaty. In order to obviate the need for inventors to deposit their organism in a culture collection in every country in which they intend to seek patent protection, the 'Budapest Treaty on the International Recognition of the Deposit of Microorganisms for the Purposes of Patent Procedure' was concluded in 1977 and came into force towards the end of 1980 (Budapest Treaty and Regulations, 1981). Under the Budapest Treaty certain culture collections are recognised as 'International Depositary Authorities' (IDAs), and a single deposit made in any one of them is acceptable by each country party to the Treaty as meeting the deposit requirements of its own national laws. Any culture collection can become an IDA provided that it has been formally nominated by a contracting state, which must also provide assurances that the collection can comply with the requirements of the Treaty. At 1 January 1987 there were 14 IDAs and they and the kinds of organisms they accept are listed in Table 6.5.

The Budapest Treaty provides an internationally uniform system of deposit and lays down the procedures which depositor and depository must follow (see Section 6.5.5 below), the duration of deposit (at least 30 years or 5 years after the most recent request for a culture, whichever is later), and the mechanisms for the release of samples. The Treaty does not, however, concern itself with the *timing* of deposit nor, in the main, of release; these are determined by the relevant national laws. Likewise, the recipients of samples (other than patent offices and people with the depositor's authorisation) are referred to merely as 'parties legally entitled': exactly *who* such parties are and under what conditions they may obtain samples are again determined by national law. Twenty one states and the European Patent Office (EPO) are now party to the Budapest Treaty and are listed in Table 6.2.

National deposit requirements. The requirements of various countries for the deposit and release of microorganisms for patent purposes are summarised in Table 6.3. Deposit is a statutory requirement under Rule 28 of the EPC and under the national laws of several of its member countries. In those EPC countries not having a statutory provision under their national law, deposit is such an established requirement of patent offices that it amounts to the same thing for all practical purposes. Most of these countries follow EPC practice in requiring the deposit to be made by the filing or priority date. An exception to this is the Netherlands, where deposit is required before the second publication (see next section) of the patent application. Most other European countries do not have specific requirements as yet, but nevertheless advise that deposits should be made, usually along the lines of the EPC.

In many countries outside Europe, deposit is an established or recommended practice, and some patent offices (in Japan, USA, USSR, for example) have specific requirements for deposit. In almost all cases deposit must be made by the filing or priority date. In the USA, however, as a consequence of a recent court decision (*in re* Lundak, 1985), deposit may in certain circumstances be made after filing but before the issuance of a US patent.

All countries party to the Budapest Treaty (Table 6.2) must recognise a

Table 6.2. *Countries party to the Budapest Treaty at 31 July 1987*

Australia	Liechtenstein
Austria	Netherlands
Belgium	Norway
Bulgaria	Philippines
Denmark	Spain
Finland	Sweden
France	Switzerland
Germany (Federal Republic)	UK
Italy	USA
Hungary	USSR
Japan	European Patent Office (EPO)[a]

[a] The EPO is not, strictly speaking, a party to the Treaty, since it is not a country but an intergovernmental organisation. Article 9 of the Treaty provides for such organisations to file a declaration stating that they accept the obligations and provisions of the Treaty. The EPO has filed such a declaration.

deposit made in an IDA but not all *require* deposits to be made in IDAs. Thus, for example, France, Germany, Switzerland, the UK, the USA and the EPC will recognise other culture collections that can comply with their particular requirements. Hungary accepts deposits made in collections on its own soil, but the only deposits it will recognise elsewhere are those made in IDAs. The Japanese patent office, however, will recognise deposits outside Japan only if they have been made under the Budapest Treaty or have been 'converted' to Budapest Treaty deposits (see 'Converted' deposits in Section 6.5.5 below), regardless of their previous public availability. It must be remembered that deposits made under the Budapest Treaty can only be made in IDAs.

Most countries not party to the Budapest Treaty accept deposits made in any internationally known culture collection which will comply with their requirements; in some cases the collection is required to furnish a declaration as to the permanence and availability of the deposit.

Release of samples. Microorganisms deposited to comply with requirements for disclosure must become available to the public at some stage of the patenting procedure. Unlike a written description, however, the microorganism is the physical essence of the invention itself and because of this the exact conditions of release are a matter of great concern to patent practitioners. There are as yet no internationally uniform release conditions, but three main kinds of system operate at present.

In the USA, patent applications are not published until the patent is granted, and a microorganism deposited for patent purposes need not be made available until then. From the date of grant, the organism must be publicly available without any restriction. The major advantage of this system to the inventor is that his microorganism does not have to be released until he has an enforceable right. Furthermore, if he is *not* granted a patent then his microorganism need never become available. Thus under the US system an inventor is never put in the position of having to allow access to his organism when he has no legal protection.

In Japan and the Netherlands patent applications are published twice, first 18 months after the filing date or priority date and before the application has been examined (Section 6.5.4 below), and second (for the purposes of opposition by third parties) when the application has been accepted. A microorganism deposited in connection with the application must be made available at the date of second publication, i.e. once the patent office has decided to grant a patent. Thus under the

Table 6.3. *National requirements (mandatory or recommended) for deposit and release of microorganisms*

Country	Deposit by	Earliest release	Earliest general availability[a]	Restrictions on distribution and use of samples	Minimum storage period (years)
Australia	F/P	1st pub.	1st pub.	as for UK	30
Austria	F/P	–	–	–	–
Belgium	F/P	1st pub.	grant	as for EPO	30
Bulgaria	F/P	grant	grant	as for UK	–
Canada	F/P	grant	grant	none	life of patent
Denmark	F/P	1st pub.	grant	as for EPO	30
Finland	F/P	1st pub.	grant	as for EPO	30
France	F/P	1st pub.	grant	as for EPO	30
Germany	F/P	1st pub.	1st pub.	until patent expires, sample must not be passed to 3rd parties or outside purview of German law	20
Hungary	F/P	1st pub.	1st pub.	sample must not be passed to 3rd party	20
Ireland	F/P	–	–	–	–
Italy	F/P	?	?	as for EPO	?
Japan	F/P	2nd pub.	2nd pub.	sample must not be passed to 3rd parties until patent expires and must be used only for research purposes	life of patent

Table 6.3. (*cont.*)

Country	Deposit by	Earliest release	Earliest general availability[a]	Restrictions on distribution and use of samples	Minimum storage period (years)
Liechtenstein				as for Switzerland	
Netherlands	2nd pub.	2nd pub.	2nd pub.	none	life of patent
New Zealand	F/P	–	–	–	–
Norway	F/P	1st pub.	grant	as for EPO	30
Portugal	F/P	–	–	–	–
Spain	F/P	1st pub.	1st pub.	–	–
Sweden	F/P	1st pub.	grant	as for EPO	30
Switzerland	F/P	?	?	sample must not be passed to 3rd parties	30
UK	F/P	1st pub.	1st pub.	sample must not be passed to 3rd parties until patent expires and must be used only for experimental purposes	30
USA[b]	F/P	grant	grant	none	30
USSR	–	–	–	–	–
EPO	F/P	1st pub.	grant	if applicant chooses, available only to independent expert before grant; must not be passed to 3rd parties before patent expires and must be used only for experimental purposes	30

F/P, filing or priority date as applicable; pub., publication of application; –, no provisions or provisions not known; ?, conflicting information from different sources.

[a] General availability means sample publicly available at least in country where application has been filed.

[b] The Lundak decision (1985) may mean that in certain cases deposit may be made later.

Japanese and Dutch systems the inventor again has an enforceable right at the time he is required to make his organism available. In Japan he is afforded a further measure of protection in that recipients of cultures must not pass them on to third parties and must use them only for experimental purposes. This provision does not apply under Dutch law, however.

A dual publication system is also operated by the EPC and by the countries party to it. However, in contrast to the Japanese requirements, a microorganism deposited in connection with an EPC application must be made available at the date of *first* publication, i.e. before any enforceable right exists. This practice reflects the prevailing philosophy of European patent authorities that the organism is regarded as an integral part of the disclosure and therefore should become available at the same time as the written description. Originally cultures had to be available to anyone requesting them, subject to the recipient giving certain undertakings of rather doubtful value, but in response to pressure from users of the system the appropriate rule (Rule 28) of the EPC was amended to provide more protection for the inventor. Rule 28(4) now permits the applicant to opt to restrict the availability of his organism at first publication to an independent expert acting on behalf of a third party. The expert, who is chosen by the requesting party from a list held by the EPO, is not permitted to pass cultures of the strain to anyone else. After second publication, the strain becomes generally available, but at this stage the inventor has an enforceable right.

Although this so-called 'expert solution' applies in respect of applications filed with the EPO itself, it is at present part of the national law of only a minority of member countries of the EPC (France, Italy, Sweden). None, however, permits recipients of cultures to pass them on to third parties.

Need for deposit. In principle most countries require deposit only when repeatability of the invention cannot be ensured without it. Thus, for example, it should not be necessary to deposit a new recombinant strain if the procedure for constructing the novel plasmid and transforming it into a host can be described in sufficient detail to allow an expert to produce the same recombinant for himself (given, of course, that the original vector and host are already generally available). In practice, however, applicants in such cases sometimes choose to deposit in order to avoid all risk of their application being rejected on the grounds of insufficient disclosure. Some applicants, on the other hand, prefer to take this risk.

In cases where the microorganism is already generally available from a culture collection, the situation is perhaps more straightforward. Some countries (e.g. Germany, USA, USSR) require the applicant to furnish a declaration signed by the culture collection and stating that the organism in question is in fact available and will remain available for the period dictated by the relevant national law. In this connection, it is worth noting that the USA in some cases presently requires such a declaration – at least for deposits made outside the USA – even where an organism has been deposited under the Budapest Treaty. As mentioned earlier, the Japanese patent office will recognise the availability of strains from culture collections outside Japan only if they have been deposited under the Budapest Treaty.

6.5   Practical considerations

So far this chapter has been concerned with the principles of biotechnological patents and the requirements of various countries; now the essentially practical aspects of the patenting process must be considered. To use specific examples, either actual or hypothetical, for this purpose would give too narrow a picture. Therefore, the more general question of seeking patent protection for an unspecified invention involving the use of a new (i.e. not already generally available) microorganism will be considered.

6.5.1   *The patent agent (patent attorney)*

The job of the patent agent is, put simply, to obtain a patent on behalf of the applicant. The agent's knowledge and understanding of patent procedures world-wide are essential to guide the applicant through the complex business of seeking patent protection, helping to draft the technical description, formulating the claims, dealing with the patent authorities, ensuring deadlines are met and so on. However, his technical knowledge of the invention and its background cannot be expected to equal that of the inventor, who must therefore be prepared to spend time and effort in familiarising him with every aspect. In his turn, the agent can often offer valuable advice on areas where more experimental work might be done before filing in order to make the application as strong as possible. The importance of these considerations is shown by the fact that many large firms have full-time patent agents on their staff to ensure that their inventions are adequately protected. The small organisation and the academic inventor, therefore, are well advised to obtain the services of a professionally qualified patent agent if they are considering applying for patent protection.

6.5.2 *Disclosing the invention*

Premature disclosure. As mentioned earlier, making a full disclosure of an invention is the applicant's side of the bargain that will give him a legal monopoly, and is a fundamental prerequisite for obtaining a patent. However, the disclosure must be made in the proper way and at the right time. Above all the invention must not be disclosed prematurely, for its novelty (see Section 6.2.2 above) will be assessed by most patent offices in the light of what is already known (the state of the art) on the day the patent application is filed. The prevailing state of the art includes any contributions the applicant himself may have made to it, whether orally, by visual display, or by display or sale of a product. Thus to avoid premature disclosure, the academic inventor in particular must abjure the normal practices of discussing his findings with other workers or publishing them in scientific journals until he has filed his patent application. Information revealed by breach of the applicant's confidence does not jeopardise a patent application, but since breach of confidence is often difficult to prove, the wisest course is to make no disclosure until the application has been safely filed. These strictures do not wholly apply (the exact conditions vary) in relation to those few countries, notably the USA, that allow a 'grace period' (see Section 6.2.2 above). However, even here it must be remembered that although the relevant national law may allow disclosure during a grace period preceding the basic national filing, such a disclosure will be prejudicial to subsequent foreign filings.

The patent specification – technical description. The patent specification contains the written disclosure or technical description of the invention and the patent claims, which state the scope and kinds of monopoly being asked for. The precise wording of the patent specification is of great importance and it is here that the skill of the patent agent comes into play in ensuring that the description (supplemented by deposit – see Section 6.5.5 below) fulfils the requirements of disclosure and that the claims are drafted to afford the best protection.

The technical description is exactly that; the invention is described in detail in scientific and technical terms – it is not addressed to the layman – and put into the context of the field to which it applies, the problems it aims to solve and the way in which solutions are achieved. The preferred format of the description is well established and typically includes the following: field, background, object and summary of the invention, followed by the detailed description of the invention. Crespi

(1982) has discussed the layout of the technical description more fully, with actual examples, and the reader is referred to his book for more information. As far as the present account is concerned, the first point is that the description must clearly convey the novelty, inventiveness and industrial applicability of the invention, and must describe the methodology in sufficient detail (worked examples being usual, but not mandatory) to enable a skilled person to reproduce the invention for himself and show that it works in accordance with the claims of the inventor. Second, the technical description must also describe any new microorganism involved in the invention. Clearly the accession number assigned to the organism by the culture collection in which it has been deposited must be quoted, but beyond that the extent of characterisation required varies between countries. The most extensive requirements are those of the Japanese patent office, which gives in its 'examination standard' a detailed list of the properties which should be recorded. The EPO has less stringent guidelines, and at the other extreme the Netherlands will accept deposit of the organism in lieu of any characterisation. Many countries expect the kind of taxonomic data that would be used in scientific publications, although most do not insist on it and accept deposit as a means of offsetting deficiencies in the written characterisation. In general, an applicant is well advised to provide characterisation data 'to the extent available to him' (Crespi, 1985).

The patent specification – claims. The claims are perhaps the most important part of the patent specification as far as the applicant is concerned, because they set out precisely the extent of the protection being sought. This is particularly so in, for example, the UK and USA where great attention is paid to the exact wording of the claims. Any loophole left here can leave the inventor exposed to competition from which he might otherwise have been protected. In Germany, on the other hand, the claims are viewed less literally in that more attention is given to them as indicators of the basic inventive idea. The EPO adopts a middle course, trying to strike a balance between the rights of the inventor and those of third parties. Again, the reader is referred to Crespi (1982) and to Ruffles (1986) for more detailed discussion about patent claims; only the salient points will be given here.

Normal practice is to make the scope of the initial claims as broad as possible, leaving it to the patent office to object if it believes that too much is being claimed for the invention. In the final event, the applicant may feel that the degree of protection he has been allowed is less than it ought to be, but this is better than finding that he is the author of his own misfortune by having claimed too little in the first place. Nevertheless, the claims must not be extravagant; they must be based on the description and be supported by it. Thus the greater the degree of novelty and ingenuity indicated by the description, and the wider the variety of worked examples given, then the broader the claims that are likely to be accepted.

Claims are usually presented as a numbered set. The first is often the broadest and is the general claim; this is followed by subclaims, each defining by example particular aspects of the general claim, and each generally being narower in scope than the one before it. The subclaims represent fall-back positions if the broader claims are held invalid. Claims of more than one kind should be included wherever possible, e.g. a new chemical compound, a microbiological process of producing it, the organism used in the process claimed *per se*, a method of diagnosis using the new compound, and a kit incorporating the new compound and conventional reagents for the diagnosis. Table 6.4 gives examples of sets of patent claims relating to different cell types.

Table 6.4. *Examples of sets of patent claims*

US Patent no. 4,567,146
　We claim:
　1. A recombinant plasmid characterized in that it contains DNA of (1) a first Rhizobium plasmid identifiable as being the same as the plasmid pVW5JI or pVW3JI of lower molecular weight present in the culture of the strain of *Rhizobium leguminosarum* NCIB 11685 or 11683 respectively and (2) a second Rhizobium plasmid found in bacteria of another strain of *Rhizobium leguminosarum,* said second plasmid having Rhizobium genes coding for nodulation, nitrogen fixation and hydrogen uptake ability but which is non-transmissible.
　2. A method of preparing a culture of bacteria of the genus Rhizobium , which method is characterized in that
　(1) in a first cross, a donor strain of Rhizobium, containing (a) a Rhizobium plasmid lacking genes coding for nodulation but which is transmissible, is crossed with a recipient strain of Rhizobium, carrying (b) a Rhizobium plasmid having Rhizobium genes coding for nodulation, nitrogen fixation and hydrogen uptake ability but which is non-transmissible, whereby a transconjugant strain carrying a plasmid which is formed from said plasmids (a) and (b) and is a conjugal precursor of a recombinant plasmid (c) having genes coding for nodulation, nitrogen fixation and hydrogen uptake ability and being transmissible is obtained;
　(2) said transconjugant strain is separated from donor and recipient strains and cultured to produce a substantially pure culture thereof;

Table 6.4. (*cont.*)

(3) in a second cross, the transconjugant strain from the first cross is used as a donor strain and crossed with a plasmid-containing recipient strain whereby a transconjugant strain carrying a recombinant plasmid (c) is obtained; and

(4) said transconjugant strain from the second cross is separated from donor and recipient strains and cultured to produce a substantially pure culture thereof.

3. A method according to claim 2 characterized in that the transmissible plasmid (a) carries at least one drug-resistance gene.

4. A method according to claim 3 characterized in that the transmissible plasmid is pVW5JI or pVW3JI, identifiable as being the same as the plasmid of lower molecular weight present in the culture of a strain of *Rhizobium leguminosarum* NCIB 11685 (pVW5JI) or NCIB 11683 (pVW3JI), and a kanamycin-resistant transconjugant strain is separated in each cross.

5. A method according to claim 2 characterized in that the transmissible plasmid (a) contains a selectable determinant.

6. A method according to claim 2, characterized in that the donor and recipient strain are of the species *Rhizobium leguminosarum*.

7. A method of impairing hydrogen uptake ability to bacteria of the genus Rhizobium, which method is characterized in that (1) a strain of *Rhizobium leguminosarum* NCIB 11684 or NCIB 11682, as a donor strain, is crossed with a recipient strain of *Rhizobium leguminosarum* to produce a kanamycin-resistant transconjugant strain, said recipient strain being one which permits selection of the transconjugant strain against the donor and recipient strains and which allows the transconjugant strain to be selected against when used as a donor in a subsequent cross with another strain of *Rhizobium leguminosarum*, (2) said transconjugant strain is separated from the donor and recipient strains and cultured to produce a substantially pure culture thereof; (3) in a second cross the transconjugant strain obtained from the first cross is used as a donor strain and crossed with a recipient strain of *Rhizobium leguminosarum* to produce a kanamycin-resistant transconjugant strain and (4) said transconjugant strain from the second cross is separated from the donor and recipient strains to produce a biologically pure culture thereof.

8. A method according to claim 7 characterized in that the recipient strain for the first cross is auxotrophic and has resistance to a drug other than kanamycin.

9. A method according to claim 7 or 8 wherein the recipient strain for the second cross is a naturally occurring strain.

10. A Rhizobium plasmid pIJ1008 having Rhizobium genes coding for streptomycin and kanamycin resistance, nodulation, nitrogen fixation and hydrogen uptake properties, which is transmissible and which is the plasmid of lowest molecular weight present in the culture of a strain of *Rhizobium leguminosarum* NCIB 11684 by virtue of the fact that it migrates the fastest on agarose gel in a gel electrophoresis determination in which a gel of 0.7% agarose in Tris-borate buffer of pH. 8.3 is subjected to electrophoresis at 25 mA and 100 volts at 4° C for 16 to 20 hours in the dark.

11. A Rhizobium plasmid pIJ1007 having Rhizobium genes coding for streptomycin and kanamycin resistance, nodulation, nitrogen fixation and hydrogen uptake properties, which is transmissible and which is the plasmid of lowest molecular weight present in the culture of a strain of *Rhizobium leguminosarum*

Table 6.4. (*cont.*)

NCIB 11682 by virtue of the fact that it migrates the fastest on agarose gel in a gel electrophoresis determination in which a gel of 0.7% agarose in Tri-borate buffer of pH 8.3 is subjected to electrophoresis at 25 mA and 100 volts at 4° C for 16 to 20 hours in the dark.

12. A biologically pure culture of bacteria of the genus Rhizobium characterized in that it contains a plasmid selected from the group consisting of pIJ1008 and pIJ1007.

13. A culture according to claim 12 of bacteria of the species *Rhizobium leguminosarum*.

14. A biologically pure culture of bacteria of the genus Rhizobium containing a recombinant plasmid characterized in that said plasmid contains DNA of (1) a first Rhizobium plasmid identifiable as being the same as the plasmid pVW5JI or lower molecular weight present in the culture of the strain *Rhizobium leguminosarum* NCIB 11685 or 11683 respectively and (2) a second Rhizobium plasmid found in bacteria of another strain of *Rhizobium leguminosarum*, said second plasmid having Rhizobium genes coding for nodulation, nitrogen fixation and hydrogen uptake ability but which is non-transmissible.

US Patent no. 4,546,082
What is claimed is:

1. A DNA expression vector capable of expressing in yeast cells a product which is secreted from said yeast cells, said vector comprising at least a segment of alpha-factor precursor gene and at least one segment encoding a polypeptide.

2. A DNA expression vector according to claim 1 wherein said segment encoding a polypeptide is an insertion into said alpha-factor precursor gene.

3. A DNA expression vector according to claim 1 wherein said segment encoding a polypeptide is a fusion at a terminus of said alpha-factor precursor gene.

4. A DNA expression vector according to claims 2 or 3 wherein coding sequences for mature alpha-factor are absent from said segment of alpha-factor precursor.

5. A DNA expression vector according to claim 1 wherein said polypeptide is somatostatin.

6. A DNA expression vector according to claim 1 wherein said polypeptide is ACTH.

7. A DNA expression vector according to claim 1 wherein said polypeptide is an enkephalin.

8. A yeast strain transformed with a DNA expression vector of claim 1.

9. A method for producing a DNA expression vector containing alpha-factor gene comprising the steps of
   (a) transforming a <u>MAT</u> alpha <u>2</u> <u>leu</u> <u>2</u> yeast strain with a gene bank constructed in plasmid YEp13;
   (b) selecting for leu transformants from the population formed in step (a);
   (c) replacing the transformants from step (b) and
   (d) screening for alpha-factor producing colonies.

10. A DNA expression vector formed according to the method of claim 9.

UK Patent no. 1,346,051
What we claim is:

1. *Fusarium graminearum* Schwabe deposited with the Commonwealth Myco-

Table 6.4. (*cont.*)

logical Institute and assigned the number I.M.I. 145425 and variants and mutants thereof.

2. *Fusarium graminearum* Schwabe I–7 deposited with the Commonwealth Mycological Institute and assigned the number I.M.I. 154209.

3. *Fusarium graminearum* Schwabe I–8 deposited with the Commonwealth Mycological Institute and assigned the number I.M.I. 154211.

4. *Fusarium graminearum* Schwabe I–9 deposited with the Commonwealth Mycological Institute and assigned the number I.M.I. 154212.

5. *Fusarium graminearum* Schwabe I–15 deposited with the Commonwealth Mycological Institute and assigned the number I.M.I. 154213.

6. *Fusarium graminearum* Schwabe I–16 deposited with the Commonwealth Mycological Institute and assigned the number I.M.I. 154210.

7. Fungal cultures containing a strain of *Fusarium graminearum* Schwabe I.M.I. 145425 or a mutant or variant thereof in a culture medium in which this strain is present in a culture medium containing or being supplied with nutrients or additives necessary for the sustenance and multiplication of the strain, the medium having a pH between 3.5 and 7 and the temperature of the medium being maintained at a precise value within the range of between 25 and 34° C.

8. A method for cultivating a strain of *Fusarium graminearum* Schwabe I.M.I. 145425 or a mutant or variant thereof wherein the strain is present in a culture medium containing or being supplied with nutrients or additives necessary for the sustenance and multiplication of the strain, the medium having a pH between 3.5 and 7 and the temperature of the medium being maintained at a precise value within the range of between 25 and 34° C.

9. A method for the preparation of variants of *Fusarium graminearum* Schwabe I.M.I. 145425 which comprises growing the parent strain I.M.I. 145425 under continuous culture conditions with carbon limitation in a fermentation.

10. A method for the preparation of variants of *Fusarium graminearum* Schwabe I.M.I. 145425 which comprises growing the parent strain I.M.I. 145425 on a glucose based medium at 25° to 30° C under continuous culture conditions at a dilution rate of 0.10 to 0.15 hrs. $^{-1}$ with carbon limitation in a fermentation for 1100 hours.

11. A method as claimed in claim 10 wherein the resulting proliferated variants are isolated by dilution plating.

12. A method for the preparation of variants of *Fusarium graminearum* Schwabe I.M.I. 145425 substantially as described with reference to Examples 1 to 5 hereinbefore set forth.

13. Fungal cultures containing *Fusarium graminearum* Schwabe I.M.I. 145425 or mutants or variants thereof substantially as described with reference to any one of Examples 6 to 11 hereinbefore set forth.

UK Patent no. 1,300,391

What we claim is:

1. A human embryo liver cell line having the characteristics of cells deposited with the American Type Culture Collection under number CL99.

2. A cell culture system comprising cells derived from the human embryo liver cell line designated by A.T.C.C. number CL99 in a nutrient culture medium therefore.

3. A virus culture system comprising cells derived from the human embryo

Table 6.4. (*cont.*)

liver cell line designated by A.T.C.C number CL99 inoculated with a virus capable of replication in said cells, and a nutrient culture medium adapted to support growth of the virus-cell system.

4. A culture system according to claim 2 or 3, wherin the nutrient culture medium contains Eagle's minimum essential medium and heat-inactivated foetal calf serum.

5. A culture system according to claim 4, wherein the nutrient culture medium contains Eagle's minimum essential medium, heat-inactivated foetal calf serum, sodium bicarbonate and one or more antibiotics.

6. A virus cultivation process which comprises maintaining a viable culture of cells derived from the human embryo liver cell line designated by A.T.C.C. number CL99 in a nutrient culture medium, inoculating the culture with a virus to which the cells are susceptible and cultivating the virus in the culture.

7. A process according to claim 6, wherein the virus is of the group consisting of adenoviruses, San Carlos viruses, ECHO viruses, arthropodborne group A viruses and other arboviruses, pox viruses, myxoviruses, paramyxoviruses, picornaviruses, herpes viruses and the AR 17 haemovirus.

8. A process according to claim 6, wherein the virus is a hepatitis virus.

9. A process according to claim 7, wherein the virus is of the group consisting of adenovirus types 2, 3, 4, 5, 7 and 17, San Carlos virus types 3, 6, 8 and 49, ECHO virus type 11, Sindbis virus, vaccinia virus, influenza A2 virus and other influenza viruses, Sabin poliovirus type 1 and other poliomyelitis viruses, and the AR—17 haemovirus.

10. A virus whenever cultivated by the process of any of claims 6 to 9.

11. Antigenic material obtained from a virus according to claim 10.

12. A vaccine comprising a virus according to claim 10, in an administrable form and dosage.

13. A vaccine comprising antigenic material according to claim 11, in an administrable form and dosage.

14. A vaccine comprising antibodies produced by a virus according to claim 10 or antigenic material according to claim 11, in an administrable form and dosage.

15. A cell culture system substantially as described in Example 3.

16. A virus cultivation process substantially as described in Example 3.

6.5.3 *Filing the application*

A single application filed in one country will result in patent protection only in that country. Therefore when the disclosure of an invention is likely to lead to serious foreign competition the normal course is to seek protection in several countries. Fortunately this does not have to be done all at once and the first application is usually filed in the applicant's own country. The patent office gives this application a number and, more importantly, a filing date. The significance of this first filing date ('priority date') is that it establishes the priority of the invention; in other words any later applications made in that country by other people for the same invention are pre-empted by it. Furthermore, provided that the applicant files any corresponding foreign applications within 12 months of his basic national filing, the original priority date is also recognised by nearly all overseas countries. However, the same priority date cannot be claimed for material not included in the basic national application (the 'priority document').

The EPO also recognises the original priority date for applications filed with it within 12 months of the basic national filing. The advantage of the European system is that an application filed with the EPO results in a clutch of 'national' patents, valid in those countries party to the EPC that the applicant has designated as being territories in which he wants patent protection. The applicant must, however, designate these countries at the outset; he can drop some of them later by not paying renewal fees, but he cannot add to them. Use of the EPC system is not mandatory in Europe however; if an inventor wishes, he can instead file separate national applications in individual countries. In fact, if protection is not required in more than two or three European countries, the national route may be cheaper.

The first steps along the road to obtaining patent protection involve drafting the patent specification, filing the basic national application and, within 12 months, filing the appropriate foreign applications (redrafting and developing further the original specification if appropriate). As Crespi (1982) has pointed out, a year is not long when account is taken of the need to evaluate the importance of the invention, decide on the extent of foreign patenting desired and implement the decision. Implementation involves drafting the final specification, sending documents around the world and possibly having the specification translated into other languages. In this last respect it is all too easy, say, for the English-speaking applicant to forget that the German Patent Office will expect an application to be written in German. (When filing with the EPO, however, the application may be drafted in English, French or German.) Thus although the patent agent will take care of all the documentary procedures, consulting the applicant where necessary, the latter cannot afford to relax completely at this stage. Attention must also be paid, of course, to ensuring that by the filing date (or, where applicable, the priority date) the microorganism used in the invention has been deposited in a suitable culture collection. This will be discussed in considerable detail later (see Section 6.5.5 below); for the present it is more convenient to follow the progress of the patent application itself.

6.5.4 *Patent office procedures*

Once the final national and/or foreign applications have been filed and the microorganism deposited, the applicant must wait for his application to be processed by the patent office. The exact procedures and the time they take vary widely between countries. Therefore, only a broad outline, illustrated with a few examples, can be given here.

All major patent offices carry out a novelty search, which is usually a patent and literature search, followed by a critical or substantive examination of the application. Under the EPC and many European national systems, the search and examination are treated separately. After the search, a report is sent to the patent agent, pointing out material (including earlier patents) considered relevant to the application. Under the EPC, this report and the patent application itself should be published 18 months after the priority (basic national filing) date, although in practice, delays in the issuance of the search report are common. At this point, the claims can be modified by the applicant if it seems unlikely that they will be accepted as they stand. Also with the publication of the application, the deposited microorganism becomes available (to varying extents) under most European systems (see Release of samples in Section 6.4.2 above). For the application to proceed further, the applicant must now ask for a substantive examination to be made of it. This request must be made within a specified period or the application will lapse.

Under the Japanese system, patent applications are published after 18 months, but there is no search or examination unless and until the applicant requests it, which he must do within seven years. In Japan and the USA, both the search and substantive examination are carried out before a report is issued to the applicant.

In all systems, a written response to the patent examiner's report must be made within a certain period or the application may fail by default. There usually then follows a variable period of negotiation with the examiner ('prosecution of the application') as to how broad the final claims should be in view of the prior art. If agreement is reached, then the application is accepted. If not, it is refused and the applicant and his agent must then consider whether to pursue an appeal to a higher tribunal.

Negotiations with the patent authorities and the meeting of deadlines are usually taken care of by the patent agent. Unless questions of a highly technical nature are raised, the involvement of the inventor himself in these proceedings is generally minimal.

After an application has been accepted, in most countries the patent is then granted and published (for the first time in the USA and Canada). At this point the deposited microorganism becomes available for the first time in the USA, Canada, Japan and (unless the EPC route has been followed) the Netherlands. Under the EPC system the microorganism, available since the first publication only to an independent expert if the applicant has so opted, becomes generally available.

Major patent offices permit a period immediately after the patent application has been allowed or the patent has been granted for it to be challenged by third parties. The extent of this 'opposition' period varies considerably between countries. Thus, for instance, the EPC allows a nine month period for after-grant revocation of a European patent (before it becomes a collection of national patents). Japan allows an opposition period of three months before grant. In the USA there is no opposition procedure as such. Instead, anyone can ask for a re-examination of any patent, regardless of when it was granted, provided that he can cite pertinent prior art previously unconsidered by the Patent Office and which is sufficient to convince the Office that the issue should be re-opened. Exceptionally, a UK patent can be revoked by application to the Patent Office at any time after grant.

The length of time for which a patent lasts once it has been granted also varies between different countries. In the USA and Canada, for instance, the period is 17 years from the date of grant, regardless of how long the application has been pending the outcome of negotiations between the applicant and the patent office. In Europe, on the other hand, the term is 20 years from the application date. In Japan, it is 15 years from publication for opposition purposes, or 20 years from the application date, whichever is the shorter. Maintenance of the patent for its full term is subject to periodic renewal fees, non-payment of which will result in the patent lapsing.

6.5.5   *Depositing the microorganism*

With an invention involving the use of a new microorganism, that is, one not already available to the public, there is one other vital act to be performed at an early stage of the patenting procedure. The microorganism must be deposited in a suitable culture collection in order to complete the disclosure of the invention. Since in almost all cases the deposit must have been effected at the latest by the filing date (or, where applicable, by the priority date), it might fairly be said that, apart from drafting the specification, this is often the first practical step to be taken towards obtaining the patent. Moreover, it is a step that relies for its effective accomplishment much more on the inventor than on his agent. The latter can do little more than advise about the documentary formalities and deadlines and perhaps suggest appropriate collections. It is the inventor who knows his organism, the technical difficulties in handling it, how long is needed to grow it, and any legal constraints in respect of its pathogenicity, which might delay matters. Thus it is up to the inventor to brief his agent so that between them they can ensure that the culture collection receives the organism in good time to allow for any possible delays or mishaps.

As mentioned above, the mechanism of deposit is now regulated internationally by the Budapest Treaty, and even in countries not yet party to the Treaty, its procedures tend to be viewed as a model system. Nevertheless, for purely national purposes, deposit under the Treaty is often not necessary (see National deposit requirements in Section 6.4.2 above). However, for the international recognition of a single deposit, using the Budapest Treaty is by far the safest course of action and the following account will be concerned mainly with the Budapest Treaty system. Although the following discussion goes into some detail, much more comprehensive information is contained in the *Guide to the Deposit of Microorganisms for the Purposes of Patent Procedure* issued by the World Intellectual Property Organization (WIPO), Geneva. For convenience, the term 'depositor' will be used in this connection in preference to 'applicant' or 'inventor'. Lastly, it should be borne in mind throughout that the date of deposit is the date on which the culture collection physically *receives* the culture, rather than the date when the culture is formally accepted.

Requirements of the Budapest Treaty. Under the Budapest Treaty a deposit must be made with an International Depositary Authority (IDA) according to the provisions of Rule 6 of the Treaty. The requirements for making such a deposit are laid down in Rule 6.1(a), which requires that the culture sent to an IDA must be accompanied by a written statement, signed by the depositor and containing the following information:

> (i) an indication that the deposit is made under the Treaty and an undertaking not to withdraw it for the period specified in Rule 9.1.

The period specified in Rule 9.1 is five years after the latest request for a sample, and in any case at least 30 years. The important thing to note here is that a deposit made under the Budapest Treaty is permanent and having made it, the depositor cannot later ask for it to be cancelled, regardless of whether a patent is eventually granted. This applies even if he abandons his patent application.

> (ii)   the name and address of the depositor;
> (iii)   details of the conditions necessary for the cultivation of the microorganism, for its storage and for testing its viability and also, where a mixture of microorganisms is deposited, descriptions of the components of the mixture and at least one of the methods permitting the checking of their presence;

This requirement simply ensures that the culture collection is given enough information to enable it to handle the organism correctly. The instructions about (intentionally) mixed cultures are included so that a positive viability statement (see below) is not issued when all the components of the co-culture are not viable.

> (iv)   an identification reference (number, symbols etc.) given by the depositor to the microorganism;

The term 'identification reference' is sometimes wrongly taken to refer to a taxonomic identification, whereas it simply means 'strain designation'.

> (v)   an indication of the properties of the microorganism which are or may be dangerous to health or the environment, or an indication that the depositor is not aware of such properties.

The requirements of Rule 6.1(a) are mandatory and cannot be varied either by the depositor or by the IDA. Indeed if the depositor does not comply with them all, the IDA is obliged to ask him to do so before it can accept the deposit. The same does not apply to Rule 6.1(b), which is not really a rule at all but simply an exhortation. According to Rule 6.1(b) 'it is strongly recommended that the written statement . . . should contain the scientific description and/or proposed taxonomic designation of the deposited microorganism'.

As well as the above requirements, the Treaty permits the IDA to set certain conditions of its own (Rule 6.3(a)). These are:

> (i)   that the microorganism be deposited in the form and quantity necessary for the purpose of the Treaty and these Regulations;

Thus an IDA may require that cultures are submitted to it in a particular state, e.g. freeze-dried, in agar stabs, etc., and that a specified number of replicates is provided.

> (ii)   that a form established by such authority and duly completed by the depositor for the purposes of the administrative procedures of such authority be furnished;

This refers to the accession form (and any other form) routinely used by the culture collection.

> (iii)   that the written statement . . . be drafted in the language, or in any of the languages, specified by such authority . . .

This is an obvious requirement, permitting a Japanese depository, for example, to ask for information to be supplied to it in Japanese.

> (iv)   that the fee for storage . . . be paid;
>
> (v)   that, to the extent permitted by the applicable law, the depositor enter into a contract with such authority defining the liabilities of the depositor and the said authority.

This provides for the IDA to make the kind of contractual arrangements with the depositor that would be usual under the laws of contract of the IDA's own country. Without this provision, some culture collections would have been unwilling to become IDAs.

It is entirely up to the IDA whether it requires any or all of the above from the depositor, but if it does, then the depositor has no option but to comply. Some of the requirements of existing IDAs are summarised in Table 6.5.

These, then, are the official requirements which the depositor must meet. For its part, the IDA also must fulfil certain obligations under the Treaty. In particular it must issue to the depositor an official receipt (the contents of which are laid down by the Treaty) stating that it has received and accepted the deposit and it must, as soon as possible, test the viability of the culture deposited and issue an official statement to the depositor informing him of the result. If the culture proves not to be viable, the deposit is worthless, which can lead to major problems (see below). The IDA must also keep the deposit secret from all except those entitled to receive samples; it must maintain the deposit for the 30 or more years required by the Treaty, checking the viability 'at reasonable intervals' or at any time on the demand of the depositor; it must supply cultures to anyone entitled under the relevant patent law to receive them (provided that the IDA has been given proof of entitlement – see below); it must inform the depositor when and to whom it has released samples; it must be impartial and available to any depositor under the same conditions.

New deposits. If by some mischance a microorganism which was viable when deposited dies during storage, or if indeed for any reason the IDA can no longer supply cultures of it, then the IDA must notify the depositor immediately. The latter then has the option of replacing it (Article 4), and provided he does so within three months, the date on which the original deposit was made still stands. When making a new deposit, the depositor must (under Rule 6.2) provide the IDA with:

(1) a signed statement that he is submitting a culture of the same microorganism as deposited previously;

(2) an indication of the date on which he received notification from the IDA of its inability to supply cultures of the previous deposit;

(3) the reason he is making the new deposit;

(4) a copy of the receipt and the last positive viability statement in respect of the previous deposit;

(5) a copy of the most recent scientific description and/or taxonomic designation submitted to the IDA in respect of the previous deposit;

(6) if the new deposit is being made with a different IDA, all the indications required under Rule 6.1(a) (see above).

With regard to item (6) above, the new deposit can be made with a different IDA if the original IDA is no longer operating as such (either entirely or just in respect of that particular kind of microorganism) or if import/export regulations render the original IDA inappropriate for that particular deposit.

It must be remembered that the provisions for making a new deposit cannot be applied to a microorganism which was shown by the IDA to be non-viable when it was originally deposited. There must have been at least one positive viability statement.

'Converted' deposits. The Budapest Treaty allows (Rule 6.4(d)) for a deposit made outside its provisions to be 'converted' to a Treaty deposit (provided, of course, that the culture collection holding the deposit is an IDA). Where the microorganism was deposited before the culture collection became an IDA the date of deposit of the 'conversion' is held

Table 6.5. *International Depositary Authorities at 1 January 1987*

Summary

International Depositary Authority	Research	Service	Culture	Microorganisms accepted	Minimum no. of replicates to be provided by the depositor
Agricultural Collection (NRRL) Peoria USA				Non-pathogenic bacteria, actinomycetes, yeasts, moulds	
American Type Culture Collection (ATCC) Rockville USA				Most kinds	
Centraalbureau voor Schimmelcultures (CBS) Baarn Netherlands				Fungi, yeasts, actinomycetes, bacteria	
Collection Nationale de Cultures de Microorganismes (CNCM) Paris France				Bacteria, actinomycetes, fungi, yeasts, viruses	
Culture Collection of Algae and Protozoa (CCAP) Ambleside and Oban UK				Algae, non-pathogenic protozoa	
Culture Collection of the CAB International Mycological Institute (CMI CC) Kew UK				Non-pathogenic fungi	

Table 6.5. (cont.)

International Depositary Authority	Microorganisms accepted	Minimum no. of replicates to be provided by the depositor
Deutsche Sammlung von Mikroorganismen (DSM)[b] Göttingen Federal Republic of Germany	Non-pathogenic bacteria, actinomycetes, fungi, yeasts, phages	
European Collection of Animal Cell Cultures (ECACC) Salisbury, UK	Cell lines, animal viruses	
Fermentation Research Institute (FRI) Ibaraki-ken, Japan	Non-pathogenic fungi, yeasts, bacteria, actinomycetes	
In Vitro International Inc. (IVI) Linthicum, USA	Most kinds	
National Collection of Agricultural & Industrial Microorganisms (NCAIM) Budapest, Hungary	Non-pathogenic bacteria, fungi, yeasts	
National Collection of Industrial Bacteria (NCIB) Aberdeen, UK	Non-pathogenic bacteria, actinomycetes, phages, plasmids	
National Collection of Type Cultures (NCTC) London, UK	Pathogenic bacteria	
National Collection of Yeast Cultures (NCYC) Norwich, UK	Non-pathogenic yeasts	

Detailed information

France

Collection Nationale de Cultures de
  Micro-organismes
(CNCM)
Institut Pasteur
28 rue du Dr Roux
75724 Paris Cedex 15

Bacteria (including actinomycetes), bacteria containing plasmids; filamentous fungi and yeasts, and viruses, EXCEPT:

– cellular cultures (animal cells, including hybridomes and plant cells);
– microorganisms whose manipulation calls for physical insulation standards of P3 or P4 level, according to the information provided by the National Institutes of Health (NIH) Guidelines for Research Involving Recombinant DNA Molecules and Laboratory Safety Monograph;
– microorganisms liable to require viability testing that the CNCM is technically not able to carry out;
– mixtures of undefined and/or unidentifiable microorganisms.

The CNCM reserves the possibility of refusing any microorganism for security reasons: specific risks to human beings, animals, plants and the environment.

In the eventuality of the deposit of cultures that are not or cannot be lyophilized, the CNCM must be consulted, prior to the transmittal of the microorganism, regarding the possibilities and conditions for acceptance of the samples; however, it is advisable to make the prior consultation in all cases.

Cell lines, 12
Other organisms, 8

Table 6.5. (cont.)

International Depositary Authority	Microorganisms accepted	Minimum no. of replicates to be provided by the depositor
Federal Republic of Germany Deutsche Sammlung von Mikroorganismen (DSM)[b] Gesellschaft für Biotechnologische Forschung mbH Grisebachstr.8 3400 Göttingen	Bacteria including actinomycetes, fungi, including yeasts, bacteriophages, except any kinds pathogenic to humans or animals. Phytopathogenic kinds are accepted, EXCEPT: *Erwinia amylovora; Coniothyrium fagacearum; Endothia parasitica; Gloeosporium ampelophagum; Septoria musiva; Synchytrium endobioticum.*	2
Hungary Mezőgazdasági és Ipari Mikroorganizmusok Magyar Nemzeti Gyüjteménye (MIMNG) [National Collection of Agricultural and Industrial Microorganisms (NCAIM)] Kertészeti Egyetem, Mikrobiológiai Tanszék (Department of Microbiology, University of Horticulture) Somlói ut 14–16 H-1118 Budapest	– Bacteria (including *Streptomyces*) except obligate human pathogenic species (e.g., *Corynebacterium diphtheriae, Mycobacterium leprae, Yersinia pestis*, etc.); – Fungi, including yeasts and moulds, except some pathogens (*Blastomyces, Coccidioides, Histoplasma*, etc.), as well as certain basidiomycetous and plant pathogenic fungi which cannot be preserved reliably. Apart from the above-mentioned, the following may not, at present, be accepted for deposit: – viruses, phages, rickettsiae; – algae, protozoa; – cell lines, hybridomes.	3 or 25[a]

Japan

Fermentation Research Institute (FRI)
1-3, Higashi 1-chome
Yatabe-machi
Tsukuba-gun, Ibaraki-ken 305

Fungi, yeast, bacteria and actinomycetes, EXCEPT:
– microorganisms having properties which are or may be dangerous to health or the environment;
– microorganisms which need the physical containment level P2, P3 or P4 required for experiments, as described in the 1979 Prime Minister's Guideline for Research Involving Recombinant DNA Molecules.

5

Netherlands

Centraalbureau voor Schimmelcultures (CBS)
Oosterstraat 1
Postbus 273
NL-3740 AG Baarn

Fungi, including yeasts; actinomycetes, bacteria other than actinomycetes.

6

United Kingdom

Culture Collection of Algae and Protozoa (CCAP)
Freshwater Biological Association
Windermere Laboratory
The Ferry House
Far Sawrey
Ambleside, Cumbria LA22 0LP
and
Scottish Marine Biological Association
Dunstaffnage Marine Research Laboratory
P.O. Box 3
Oban, Argyll PA34 4AD

(i) Freshwater and terrestrial algae and free-living protozoa (Freshwater Biological Association); and
(ii) marine algae, other than large seaweeds (Scottish Marine Biological Association).

6

Table 6.5. (*cont.*)

International Depositary Authority	Microorganisms accepted	Minimum no. of replicates to be provided by the depositor
Culture Collection of the Mycological Institute (CMI CC) Ferry Lane Kew Surrey TW9 3AF	Fungal isolates, other than known human and animal pathogens and yeasts, that can be preserved without significant change to their properties by the methods of preservation in use.	6
European Collection of Animal Cell Cultures (ECACC) [formerly known as the National Collection of Animal Cell Cultures (NCACC)] Vaccine Research and Production Laboratory Public Health Laboratory Service Centre for Applied Microbiology and Research Porton Down Salisbury, Wiltshire SP4 0JG	Cell lines that can be preserved without significant change to or loss of their properties by freezing and long term storage; viruses capable of assay in tissue culture. A statement on their possible pathogenicity to man and/or animals is required at the time of deposit. Up to and including ACDP Category 3 can be accepted for deposit (Advisory Committee on Dangerous Pathogens: Categorisation of Pathogens according to Hazard and Categories of Containment ISBN 0/11/883761/3 HMSO London).	12, each containing at least 2×10^6 cells
National Collection of Industrial Bacteria (NCIB) c/o The National Collections of Industrial and Marine Bacteria Ltd. Torry Research Station P.O. Box 31 135 Abbey Road Aberdeen AB9 8DG	(a) Bacteria, including actinomycetes, that can be preserved without significant change to their properties by liquid nitrogen freezing or by freeze-drying (lyophilisation), and which are allocated to a hazard group no higher than Group 2 as defined by the UK Advisory Committee on Dangerous Pathogens (ACDP);	Bacteria and phages, 2 Naked plasmids, 20

United Kingdom (*cont.*)

(b) Plasmids, including recombinants, either

(i) cloned into a bacterial or actinomycete host, or

(ii) as naked DNA preparations.

As regards (i) above, the hazard category of the host with or without its plasmid must be no higher than ACDP Group 2.

As regards (ii), above, the phenotypic markers of the plasmid must be capable of expression in a bacterial or actinomycete host and must be readily detectable. In all cases, the physical containment requirements must not be higher than level II as defined by the UK Genetic Manipulation Advisory Group (GMAG) and the properties of the deposited material must not be changed significantly by liquid nitrogen freezing or freeze-drying.

(c) Bacteriophages that have a hazard rating and containment requirement no greater than those cited in (a) or (b), above, and which can be preserved without significant change to their properties by liquid nitrogen freezing or by freeze-drying.

Notwithstanding the foregoing, the NCIB reserves the right to refuse to accept any material for deposit which in the opinion of the Curator presents an unacceptable hazard or is technically too difficult to handle.

Table 6.5. (cont.)

International Depositary Authority	Microorganisms accepted	Minimum no. of replicates to be provided by the depositor
United Kingdom (cont.)	In exceptional circumstances the NCIB may accept deposits which can only be maintained in active culture, but acceptance of such deposits, and relevant fees, must be decided on an individual basis by prior negotiation with the prospective depositor.	
National Collection of Type Cultures (NCTC) Central Public Health Laboratory 175 Colindale Avenue London NW9 5HT	Bacteria that can be preserved without significant change to their properties by freeze-drying and which are pathogenic to man and/or animals.	1
National Collection of Yeast Cultures (NCYC) Food Research Institute Colney Lane Norwich, Norfolk NR4 7UA	Yeasts other than known pathogens that can be preserved without significant change to their properties by freeze-drying or, exceptionally, in active culture.	1
United States of America		
American Type Culture Collection (ATCC) 12301 Parklawn Drive Rockville, Maryland 20852	Algae, animal viruses, bacteria, cell lines, fungi, hybridomas, oncogenes, phages, plant tissue cultures, plant viruses, plasmids, protozoa, seeds, yeasts. The ATCC must be informed of the physical containment level required for experiments using the host vector system, as described in the 1980 National Institutes of Health Guidelines for Research involving Recombinant DNA Molecules (i.e. P1, P2, P3 or P4 facility). The ATCC, for the time being, will accept only those hosts containing	Animal viruses Cell lines } 25 Naked plasmids Other organisms, 6 Seeds, 250

Agricultural Research Service Culture Collection
(NRRL)
1815 North University Street
Peoria, Illinois 61604

plasmids which can be worked in a P1 or P2 facility.

Certain animal viruses may require viability testing in an animal host, which the ATCC may be unable to provide. In such cases, the deposit cannot be accepted. Plant viruses which cannot be mechanically inoculated also cannot be accepted.

Progeny of strains of agriculturally and industrially important bacteria, yeast, molds, and *Actinomycetales*, EXCEPT:

(a) *Actinobacillus* (all species); *Acytomyces* (anaerobic/microaerophilic – all species); *Arizona* (all species); *Bacillus anthracis*; *Bartonella* (all species); *Bordetella* (all species); *Borrelia* (all species); *Brucella* (all species); *Clostridium botulinum*; *Clostridium chauvoei*; *Clostridium haemolyticum*; *Clostridium histolyticum*; *Clostridium novyi*; *Clostridium septicum*; *Clostridium tetani*; *Corynebacterium diphtheriae*; *Corynebacterium equi*; *Corynebacterium haemolyticum*; *Corynebacterium pseudotuberculosis*; *Corynebacterium pyogenes*; *Corynebacterium renale*; *Diplococcus* (all species); *Erysipelothrix* (all species); *Escherichia coli* (all enteropathogenic types); *Francisella* (all species); *Haemophilus* (all species); *Herellea* (all species); *Klebsiella* (all species); *Leptospira* (all species); *Listeria* (all species); *Mima* (all species); *Moraxella* (all species); *Mycobacterium avium*; *Mycobacterium bovis*; *Mycobacterium tuberculosis*; *Mycoplasma* (all species); *Neisseria* (all species); *Pasteurella* (all species); *Pseudomonas pseudomallei*; *Salmonella* (all species); *Shigella* (all species);

1 or 30[a]

Table 6.5. (cont.)

International Depositary Authority	Microorganisms accepted	Minimum no. of replicates to be provided by the depositor
United States of America (cont.)	*Sphaerophorus* (all species); *Staphylococcus aureus*; *Streptobacillus* (all species); *Streptococcus* (all pathogenic species); *Treponema* (all species); *Vibrio* (all species); *Yersinia* (all species);	
	(b) *Blastomyces* (all species); *Coccidioides* (all species); *Cryptococcus* (all species); *Histoplasma* (all species); *Paracoccidioides* (all species);	
	(c) *Basidiomycetes* or other molds that cannot successfully be preserved by lyophilization (freeze-drying);	
	(d) all viral, Rickettsial, and Chlamydial agents;	
	(e) agents which may introduce or disseminate any contagious or infectious disease of animals, humans, or poultry and which would require a permit for entry and/or distribution within the United States of America;	
	(f) agents which are classified as Plant Pests and which would require a permit for entry and/or distribution within the United States of America;	
	(g) mixtures of microorganisms;	
	(h) fastidious microorganisms which would require (in the view of the Curator) more than reasonable attention in handling and preparation of lyophilized material;	

United States of America (*cont.*)

In Vitro International, Inc.
(IVI)
611(P) Hammonds Ferry Road
Linthicum, Maryland 21090

(i) phage of any kind;
(j) plasmids and like materials.

Algae, bacteria with plasmids, bacteriophages, cell cultures, fungi, protozoa and animal and plant viruses. Recombinant strains of microorganisms will also be accepted, but IVI must be notified in advance of accepting the deposit of the physical containment level required for the host vector system, as prescribed by the National Institutes of Health Guidelines. At present, IVI will accept only hosts containing recombinant plasmids that can be worked in a P1 or P2 facility.

Bacteria ⎫
Fungi   ⎬ 3
Yeasts   ⎭
Seeds, 400
Other organisms, 6

---

[a] If depositor's own lyophilised cultures are to be stored and distributed.

[b] See present address on p. 25.

to be the date on which the collection acquired IDA status. Otherwise the date of deposit is the date on which the collection physically received the culture. The procedure for converting a deposit usually involves completing the same forms as are used for making a deposit *de novo*. However, only the original depositor (or his successor) can convert a deposit. In all other cases, a separate deposit of the same organism must be made under the Treaty.

Conversion is a useful facility because it means that an earlier non-Budapest Treaty deposit can be accorded the international recognition which it might not otherwise command. Conversion is essential for the recognition by the Japanese patent office of any non-Budapest Treaty deposit made outside Japan.

Guidelines for deposit. It is generally recognised that many depositors (and sometimes their patent agents) are unlikely to be familiar with the minute details of the Budapest Treaty and may not be aware of their obligations in respect of it. Therefore, the forms which IDAs ask prospective depositors to fill in are generally so designed that by completing them correctly the depositor automatically provides all the information required of him by the Treaty. These forms vary to some extent between IDAs, but they all follow a similar general pattern. Any IDA will supply specimens of its forms on request.

Making a deposit under the Budapest Treaty should be quite straight-forward, but problems can and do arise. It has to be said that many of these are of the depositor's (or his agent's) own making and that they can be avoided by adhering to a few simple guidelines. Perhaps the most important thing to be remembered is that the Budapest Treaty procedures take a certain amount of time to complete, even when they are operating ideally. Thus although in principle a deposit does not have to reach the IDA until the filing (or priority) date of the relevant patent application, in practice the wise depositor will start the depositing procedure in good time to allow for any possible delays. Thus if he is intending to deposit in a foreign IDA, say, he should bear in mind any import or quarantine regulations. For instance, it can take several weeks, or even months, to obtain a permit to import cell lines and viruses into the USA. Last-minute deposits are unwise for several reasons, some of the most common being:

(1) postal delays: the culture fails to arrive in time;
(2) customs delays: with deposits from overseas, depositors have not provided adequate shipping information;

(3) the deposit is not the kind of microorganism accepted by the IDA (see Table 6.5);

(4) the microorganism cannot be recovered from the package, e.g. because the culture tube is broken;

(5) the deposit proves to be non-viable; if a microorganism is found by the IDA from the outset to be non-viable, the original date of deposit cannot be applied to any replacement (see above).

For most bacteria, fungi, yeasts, algae and protozoa, viability testing usually takes 3–5 days; for animal cell lines a week or slightly longer is normal; and for animal viruses and plant tissue cells, up to a month is not unusual.

It cannot be emphasised too strongly that however good the depositor's intentions may be, patent offices recognise only the actuality of the deposit. With all this in mind the prudent depositor will also pay attention to a few more elementary points to ensure a timely and trouble-free deposit. He will ensure that the microorganism he wishes to deposit is one of the kinds that the IDA he had chosen can officially accept under the Budapest Treaty (see Table 6.5). If there are likely to be technical problems with the organism he will advise the IDA in advance. He will check the administrative and technical requirements of the chosen IDA and ask for the appropriate patent deposit forms, which he will then fill in completely and correctly, for by doing so he should automatically comply with the requirements of Rule 6.1(a) (see above). Although Rule 6.1(a) states that the microorganism should be *accompanied* by a written statement (the completed deposit form), in practice it is often helpful to an IDA to receive the written information in advance of the microorganism itself, so that arrangements can be made to deal with the deposit promptly. This is particularly helpful if, say, a special growth medium has to be prepared by the IDA. Lastly, if the depositor's patent agent is likely to be communicating with the IDA, the depositor should let the IDA know, otherwise it may withhold information until it has ascertained the agent's right to receive it.

Depending on its policy and on the kind of material being deposited, an IDA may or may not prepare subcultures for eventual distribution. Thus in the case of cell lines and naked plasmids (not cloned into a host), for instance, the depositor is usually required to supply sufficient material for the IDA to distribute direct. On the other hand, for bacteria, yeasts, moulds, etc. (with or without plasmids) it is more usual for the IDA to distribute its own preparations. In this case, many IDAs will ask the depositor to check the authenticity of their preparations – a fairly normal culture collection practice. The depositor is not *obliged* by the Budapest Treaty to check these preparations, but he is well advised to do so to ensure that the cultures to be sent out by the IDA will in fact do what is claimed for them in the patent application.

The official aspects of the depositing procedure end with the issuing of the receipt and viability statement by the IDA. These are important for they are the documentary proof that on a particular date a viable deposit has been made according to the terms of the Budapest Treaty.

Technically the receipt should be issued first, but in practice many IDAs find it more convenient to await the results of the viability test and then send out the receipt and viability statement together. In general, for deposits of most bacteria, fungi, yeasts, algae and protozoa, the depositor could expect an IDA to send him both documents within a few days of it having received the deposit. For animal cell lines a week or slightly longer would be normal, and for animal viruses and plant tissue cells four or five weeks would be more usual.

6.5.6   *Obtaining a sample of a patent deposit*

So far in this account the deposit procedure has been considered primarily from the viewpoint of the depositor. It would be useful now to look briefly at the procedures which a third party must follow in order to obtain a culture, since the whole point of deposit is to make the microorganism available.

It is generally admitted that culture collections can neither be expected to be familiar with the patent laws of countries throughout the world nor to know what stage patent applications relating to the deposits they hold have reached. Thus to require a collection to judge for itself whether a particular person is legally entitled to a culture of a particular deposit is considered by many to impose an unfair burden on the collection. Therefore the Budapest Treaty attempts to place the onus on patent offices to ensure that IDAs are not put in this position (Rule 11.3). Patent offices in countries whose laws require that deposited microorganisms must be available without restriction to anyone once the relevant patents have been granted and published can notify the IDAs from time to time of the accession numbers of the strains cited in these patents. However, this provision is not usually adopted. The US Patent Office, for instance, directs that a microorganism must be available from the date of issuance of the relevant US patent, but it does not advise the IDA of this date. Since the issuance or not of a US patent is a simple matter of fact, in the event of any request an IDA merely has to ascertain this fact, either from the requesting party or the depositor. In the author's experience this does not cause any major problems, although the actual furnishing of the sample may be delayed slightly while evidence of publication is obtained. The IDA can then meet any requests for the strains in question without the need for evidence of entitlement.

In cases where the availability of the microorganism is restricted and/or where evidence of entitlement to receive a sample is required, anyone requiring a culture must either obtain the written authorisation of the depositor or he must obtain from a relevant patent office a certificate stating:

(1) that a patent application in respect of the strain in question has actually been filed with that office;

(2) whether the application has been published;

(3) that the person requesting the culture is legally entitled to receive it and has met any conditions that the law requires.

On receipt of a request accompanied by such a certificate, or by the written authorisation of the depositor, the IDA will supply the culture (subject to its normal fee for such cultures being paid). At the same time, the IDA will inform the depositor when and to whom it has supplied the culture, as it is obliged to do by Rule 11.4(g) of the Treaty, unless the depositor has specifically waived his right to be informed.

Except where the direct authorisation of the depositor has been sought, the request for a culture must be made on an official form which can be had from the patent office(s) with which the relevant application has been filed. Most IDAs also have copies of these forms. Thus, the procedure for obtaining a culture of a microorganism deposited under the Budapest Treaty is:

(1) ask the appropriate patent office, or the IDA, for a copy of the form to be used for requesting samples of microorganisms deposited under the Budapest Treaty;

(2) complete that part of the form to be filled in by 'the requesting party';

(3) send the entire form to the patent office, *not* to the IDA;

(4) when the form bearing the appropriate stamp of authorisation is received back from the patent office, send it to the IDA along with a normal purchase order.

Procedures for obtaining organisms deposited for patent purposes outside the Budapest Treaty vary according to the national law. In such cases, the culture collection will have been informed by the depositor or the patent office of the appropriate requirements and should be able to advise accordingly.

It must be remembered that the procedures outlined above relate only to the right to receive cultures according to patent law. They do not override any requirements to be met in respect of import and quarantine regulations, health and safety procedures, plant disease regulations, etc. Thus as well as obtaining patent office authorisation, a person requesting a culture must also ensure that he has obtained any permit or licence necessary for handling the organism in question.

6.6   Further reading

Adler, R. G. (1984). Biotechnology as an intellectual property. *Science* **224**, 357–63.

Anonymous (1982). Japanese Patent Office guidelines for examination of inventions of microorganisms. *Yuasa and Hara Journal* **9** (3).

Biggart, W. A. (1981). Patentability in the United States of microorganisms, processes utilizing microorganisms, products produced by microorganisms and microorganism mutational and genetic information techniques. *IDEA, Journal of Law and Technology* **22**, 113–36.

Byrne, N. J. (1979). Patents on life. *European Intellectual Property Review* **1**, 279–300.

Byrne, N. J. (1983). The agritechnical criteria in plant breeders' rights law. *Industrial Property* (1983), 294–303.

Cooper, I. P. (1982). *Biotechnology and the Law.* New York: Clark Boarman Co.

Crespi, R. S. (1981). Biotechnology and patents – past and future. *European Intellectual Property Review* **3**, 134–40.

Crespi, R. S. (1985). Biotechnology patents – a case of special pleading? *European Intellectual Property Review* **7**, 190–3.

Crespi, R. S. (1985). Microbiological inventions and the patent law – the international dimension. *Biotechnology and Genetic Engineering Reviews* **3**, 1–37.

Crespi, R. S. (1986). Patent issues in biotechnology. In *Biotechnology and Crop Improvement and Protection*, British Crop Protection Council Monograph No. 34, ed. Peter R. Day, pp. 209–17.

Halluin, A. P. (1982). Patenting the results of genetic engineering research: an overview. In *Patenting of Life Forms*, Banbury Report No. 10, pp. 67–126. Cold Spring Harbor, New York: Cold Spring Harbor Laboratory.

Hüni, A. (1977). The disclosure in patent applications for microbiological inventions. *International Review of Industrial Property and Copyright Law* **8**, 500–21.

Hüni, A. & Buss, V. (1982). Patent protection in the field of genetic engineering. *Industrial Property* (1982), 356–68.

Irons, E. S. & Sears, M. H. (1975). Patents in relation to microbiology. *Annual Review of Microbiology* **29**, 319–32.

Plant, D. W., Reimers, N. J. & Zinder, N. D. (eds.) (1982). *Patenting of Life Forms*. Banbury Report No. 10. Cold Spring Harbor, New York: Cold Spring Harbor Laboratory.

Pridham, T. G. & Hesseltine, C. W. (1975). Culture collections and patent depositions. *Advances in Applied Microbiology* **19**, 1–23.

Teschemacher, R. (1982). Patentability of microorganisms *per se*. *International Review of Industrial Property and Copyright Law* **13**, 27–41.

Wegner, H. C. (1979). Patenting the products of genetic engineering. *Biotechnology Letters* **1**, 145–50 and 193.

Wegner, H. C. (1980). The Chakrabarty decision patenting products of genetic engineering. *European Intellectual Property Review* **2**, 304–7.

The author gratefully acknowledges the helpful comments and criticisms made by many colleagues during the writing of this chapter. Particular thanks go to Mrs B. A. Brandon of the American Type Culture Collection and Mr R. S. Crespi of the British Technology Group. A singular debt of gratitude is owed to Mr R. K. Percy of the British Technology Group, for his extensive advice and painstaking correction of the original draft.

7

Culture collection services

D. ALLSOPP and F. P. SIMIONE

7.1 Introduction

In response to the needs of users, many culture collections provide a range of services to the scientific, technological and commercial world. This chapter provides an introduction to the types of services available from culture collections, but it is beyond its scope to give a comprehensive list of such services. As the range of work that can be undertaken is increasing at many of the collections, the reader should contact individual collections to find out whether they can offer particular services.

7.2 Types of services

7.2.1 Directly associated and customer services

The two major services which are intrinsically part of culture collection work are those concerning the identification and preservation of organisms. Collections of necessity need expertise in these fields to be able to function, and many provide comprehensive services in these areas. Aspects of culture identification methods (Chapter 5), sales of cultures (Chapter 3), preservation techniques (Chapter 4) and patent deposits (Chapter 6) are covered elsewhere in this volume.

Safe-Deposits. Many collections hold organisms which are not listed in their catalogues. These cultures are held for a variety of reasons: the organisms may not be fully identified, their taxonomic status may be unclear, their stability in preservation may be suspect or they may be held at the request of the depositor who wishes to have back-up material and yet retain ownership and confidentiality, not releasing the strain to other parties. Many collections have introduced safe-deposit services as a back-up to the depositor's working collection, providing a service intermediate between an open collection deposit and a deposit for patent purposes. Such services enable the depositor to have important organisms professionally preserved and maintained even if the collection would not normally be interested in accessioning them. There is obvious merit in the safekeeping of cultures while they are the subject of research, especially as many laboratories do not have optimum preservation facilities. Most collections make a charge for safe-deposits to cover the long-term storage costs and quality control procedures that are required.

Advice on strain selection. Collections are able to give advice on the selection of strains for special purposes. Such services may involve collection staff in a considerable amount of work and are limited by the sources of information available. In the past such services have relied upon the individual expertise of collection staff and their personal knowledge of the scientific literature. With the development of computer databases and strain data networks (Chapter 2), such services are becoming more frequent and are increasing in efficiency. However, as databases are searched electronically and appropriate microorganisms selected, the expertise of the collection staff is still needed to draw attention to closely allied genera and species that might merit study or to point out the idiosyncracies of individual strains. These advisory services are being placed on a more formal basis and charges may be made.

Advice on maintenance of organisms. Most collections are able to provide information on preservation systems, either on a formal or informal basis. Some of the major collections have produced substantial publications covering the preservation of their own groups of organisms and these may be consulted. However, if any doubt exists or if different media are being tried, the collections can always be consulted for advice. Advisory sheets on particular preservation techniques, or related topics such as the handling of pathogens, elimination of contamination or mite infestation, are often available on request.

Biochemical services. Some groups of organisms, such as bacteria and yeasts, are identified using biochemical tests, and such methods are increasingly used for other organisms, particularly the filamentous fungi. The need for taxonomic clarification using biochemical criteria goes hand in hand with an increasing requirement to provide metabolic and other physiological data to users of the collections. Applied biological industries, and in particular biotechnology, increasingly seek organisms on the basis of activity rather than name, and collections are responding to this requirement.

Collections now enhance their catalogues with strain data, and supply information to strain databases and computer networks as a routine procedure. Some collections also carry out custom-designed screening programmes to select strains with specified attributes for individual clients, particularly in the fields of enzyme and secondary metabolite (including toxin) production or specific growth requirements.

These biochemical services serve both as directly associated and contract services of culture collections (see also Research and development work in section below).

7.2.2   *Contract services*

The services outlined in Section 7.2.1 above usually exist as a consequence of the collection's normal activities; however, in recent years there has been an expansion in other services offered by collections for a variety of reasons. For example, many collections are associated with other institutions, such as research organisations, taxonomic institutes, educational institutions, or commercial firms, any of which may make specific demands on the services of the collection. Again, the collection may be supported by funds provided from external sources which may require particular expertise or services to be developed. Many collections are currently under financial pressure to increase their earning potential, and this has stimulated the development of income-generating activities.

Biological testing. Many modern standards and specifications require the use of microorganisms and cell lines in the testing of products. These include the mould growth resistance of materials (Fig. 7.1), the testing of disinfectants against a range of bacteria, the assessment of mutagenicity of materials against a range of organisms, and toxicity testing. For economic and other reasons there is strong pressure to move away from the use of live animals in the testing of products and towards the use of microorganisms or cell lines instead.

These tests can be carried out in any suitably equipped laboratory, but culture collections are in a very favourable position to carry out such testing themselves, using the organisms or cell lines which they would normally supply to outsiders for such work; indeed, collections may be identified in the standards themselves as an approved source of such material. Unless a company is equipped for routine testing as part of its

Fig. 7.1. Chamber used for the testing of industrial products for resistance to mould growth at the CAB International Mycological Institute (IMI).

quality control procedures, it is often more cost effective for such work to be carried out in a specialised laboratory. To establish a laboratory and train staff to carry out extensive biological testing infrequently would be very expensive, and not necessarily satisfactory from a technical point of view. However, an economic service can be offered by culture collections equipped to carry out such work on a regular basis.

Examples of testing standards using microorganisms and cell lines include British (BS), European (ISO) and USA (ASTM, USP) standards for fungal resistance testing, sterility testing, preservative effectiveness, toxicity and biocompatibility testing (Table 7.1). Culture collections provide reference cultures used in clinical laboratory standards procedures (e.g. NCCLS, ECCLS), and can provide uniform, quality assured sets of reference cultures for use in on-site biological testing for both clinical and industrial applications.

Table 7.1. *Examples of testing standards involving microorganisms, cell cultures and related materials*

British Standards	
BS 1982	Methods of testing for fungal resistance of manufactured building materials made of, or containing, materials of organic origin
BS 2011	The environmental testing of electronic components and electronic equipment. Test J. Mould growth
BS 28458	Flexible insulating sleeving for electrical purposes. Section 12. Mould growth
BS 3046	Specification for adhesives for hanging flexible wall coverings. Appendix G. Test for susceptibility to mould growth
BS 4249	Specification for paper jointing. Section 5.7. Resistance to mould growth
BS 5980	Specification for adhesives for use with ceramic tiles and mosaics. Section 7. Resistance to mould growth
BS 6009	Wood preservatives. Determination of toxic values against wood destroying *Basidiomycetes* on an agar medium
BS 6085	Methods of test for the determination of the resistance of textiles to microbiological deterioration
European Standards	
ISO 846	Plastics – determination of behaviour under the action of fungi and bacteria – evaluation or measurement of change in mass or physical properties

See also NFX41–514 and DIN 53739 for similar methods of plastics testing in France and Germany, respectively.

Table 7.1. (*cont.*)

US Standards	
ASTM D2574	Standard test method for resistance of emulsion paints in the container to attack by microorganisms
ASTM D3273	Resistance to growth of mould on the surface of interior coatings in an environmental chamber. Standard test method for evaluating the degree of surface disfigurement of paint films by fungal growth or soil and dirt contamination
ASTM G21–70	Standard recommended practice for determining resistance of synthetic polymeric materials to fungi
ASTM G22–76	Standard recommended practice for determining resistance of plastics to bacteria
FDA 21CFR610.12	Sterility testing of biological products
FDA 21CFR610.30	Detection of mycoplasma contamination
MIL STD 810D	Environmental test methods. Method 508.2 Fungus
NCCLS M2–A3	Performance standards for antimicrobial disk susceptibility tests
NCCLS M7A	Methods for dilution antimicrobial susceptibility tests for bacteria that grow aerobically
NCCLS M11A	Reference agar dilution procedure for antimicrobial susceptibility testing of anaerobic bacteria
USDA 9CFR113.26	Detection of viable bacteria and fungi in biological products
USDA 9CFR113.28	Detection of Mycoplasma contamination
USDA 9CFR113.29	Determination of moisture content in desiccated biological products
USP	Antimicrobial preservatives – effectiveness assay
USP	Microbial limits tests
USP	Sterility tests

Consultancy. Many collections have staff with expertise in specialised areas commensurate with their collection responsibilities, who can be made available for consultancy work. Such work is often a forerunner of detailed investigations, or a research programme, and can be time consuming. Since collection staff time for this work is limited, consultancy work is usually offered on a fully charged basis.

Research and development work. Culture collections equipped for testing work and consultancy may also be involved in research and development work and in industrial investigations which require laboratory facilities. It is often difficult for collections to specify exactly what kind of work they would be prepared to accept, as many problems are unique and may be carried out on a one-off basis. It is usual, therefore, for collections to consider any type of work which falls in their general area of competence. By their very nature, culture collections usually have a wide range of scientific and industrial contacts, and even if they are not able to carry out particular investigations themselves, they may be able to advise on places where such work could be carried out. Culture collections can therefore act as referral centres and this function should not be overlooked.

In addition to research topics on taxonomy and the preservation of organisms, major culture collections are often well placed to undertake other research, often allied to biotechnology, which may be initiated by short-term contracts and testing work. Potential topics for such longer-term research and development work might include:

(1) screening large numbers of isolates for particular biochemical properties and end uses;
(2) comparative studies in regard to enzyme or metabolite production of different strains of the same or closely related species;
(3) development and evaluation of rapid detection procedures for compounds produced by organisms, including the development of commercial kits;
(4) studies on the use of microorganisms as bio-control agents against insect pests, weeds and deteriogenic microorganisms;
(5) selection of test organisms used in the evaluation of materials;
(6) studies on growth requirements and the testing of bioreactors;
(7) evaluation of media, culture vessels, diagnostic reagents and procedures.

Custom preparations. Culture collections may be requested to provide bulk amounts of their organisms on an *ad hoc* basis to commercial organisations with limited facilities. Often this is an efficient and cost-effective method of obtaining bulk inocula, since the collection has the expertise in growing the organisms and ready access to the media.

Government agencies may request multiple units of mixtures of cultures for laboratory certification and proficiency testing in clinical laboratories. Manufacturers of diagnostic instruments may specify cultures with known properties for calibration of their instruments.

Resource development. In the United States the Federal Government has used culture collections extensively to develop research reagents/

cultures. These have been developed through joint efforts with the research scientists defining the needs and quality controls required and the banking and distribution aspects handled by the collections.

7.2.3 *Confidentiality*

Work performed by culture collections can be carried out on a confidential basis if required. Formal mechanisms for ensuring confidentiality are in fact required by some laboratory accreditation schemes (see below) and are already in place in a number of collections. In some areas, such as identification services, users are sometimes content for material sent in to become part of the open collection, should the collection wish to retain it. However, in collections which deal with organisms of industrial importance, it may be more normal to carry out the work in confidence, with all material being destroyed after examination. Enquiries can be treated as confidential, and collections should be questioned about their policy and procedures regarding confidentiality. Submissions are a source of new organisms for building up the resources available from the collections, and some collections will identify cultures for a lower fee if the culture can be retained in the collection.

7.2.4 *Laboratory accreditation*

In many countries national schemes exist for the accreditation of testing laboratories. Such schemes aim to establish standards for the accuracy and efficiency of measuring instruments, quality control procedures, record-keeping, and administration of the laboratory. Major companies may have their own accreditation schemes for laboratories that they use, but where national schemes exist, companies often accept the accreditation of the scheme and do not carry out individual laboratory accreditation on their own behalf. National schemes may also be recognised on an international basis and thus ease the way for laboratories to be accepted on a much wider geographical basis. There are now a sufficient number of national and international testing standards involving microorganisms to make such accreditation schemes worthwhile for culture collection laboratories to adopt.

7.3 **Workshops and training**

Many culture collection staff are involved in educational activities, either because of their general scientific background or their detailed knowledge of culture collection practices. Instruction may be external, where staff are involved in training courses at local colleges, polytechnics or research institutes in their own country or overseas. Increasingly, however, collections are devising formal programmes of in-house courses. These may be arranged in direct response to a need identified either by the collection itself or by an outside body. In some parts of the world the educational system is such that credits towards a qualification can be gained by attendance at approved outside courses, and where such a system exists, it can ease the task of the culture collection in operating such courses.

Courses vary in length from several weeks' or even months' duration (if dealing with topics such as identification), to one-day lecture courses or seminars on specific industrial or commercial topics. Outside speakers and instructors may be used to enable topics to be covered which extend beyond the expertise available in the collection itself. In addition to formal programmed and advertised courses, it is often possible to obtain individual training within culture collections to suit a particular need. Some collections have special facilities and staff devoted to training programmes, while others may use external facilities at educational institutions nearby. Charges are usually made for training, but in some cases the costs may be subsidised. Advice may be available from collections on suitable sources of funding for prospective students and trainees, especially from developing countries. Some culture collections are either owned by academic institutions such as universities, or are officially associated with them, enabling them to offer training towards MSc and PhD degrees by research.

Some of the training offered is directly related to the normal activities of the collection, and instruction in preservation and maintenance techniques is often available; indeed, such training is often difficult to obtain from other sources. The training facilities offered by culture collections are often much greater than is generally known and the World Federation for Culture Collection's Education Committee is developing a list of teachers, their special expertise and courses available throughout the world (see Chapter 8). Individual collections may also provide information on training facilities in their scientific speciality or geographical area.

7.4   Publications, catalogues and publicity material

The essential publication of any culture collection is its catalogue. Traditionally, these have been produced as hard-copy items but are now becoming available in a computerised on-line form (see Chapter

2). It is always worthwhile contacting a collection if an organism does not appear in its most recent hard-copy catalogue, as additional material may well be available, or information can be given on suitable organisms in reserve collections, which could possibly be released. In addition to catalogues, a few of the major collections produce scientific publications of their own (guides on preservation and maintenance, safety and handling, industrial uses and teaching) as well as articles in the scientific press and monographic studies. Details of such publications may be found in the collection's brochures or newsletters, or in bibliographic databases.

Collection brochures are generally available free of charge, and it is worthwhile asking to be included on the collection's mailing list to ensure receipt of up-to-date information. Despite the fact that collections grow and change with time, enquirers or customers often rely on data from back issues of catalogues which may be many years old. This can lead to considerable confusion and users should ensure that they have the current catalogues before ordering cultures. These problems are minimised as catalogue and strain data become widely available through computer networks. For many people, however, printed information remains the most important reference material.

7.5   Fees and charges

The ways in which culture collections are funded are extremely diverse. Very few culture collections exist as straightforward commercial entities; they are almost all subsidised in some fashion, either directly or indirectly, and the charges made for cultures do not reflect the true cost of production. Nevertheless, many collections offer discounts for bulk orders, special sets, or regularly ordered organisms such as those used for testing and teaching.

The charges made for services, however, more accurately reflect the true cost, though they usually represent good value in comparison with totally commercial services, particularly in the areas of training. Charges for consultancy work, for testing and for laboratory services are normally at competitive commercial rates.

Over the last few years, several important stimuli have been applied to culture collections, including the advent of biotechnology, the development of computerised databases and a harsher economic climate. These and other factors have led culture collections to examine the services they provide and to develop them to cater to the changing demand of their users. Expansion and diversification of the range of services has resulted.

7.6   Suggested reading

Alexander, M., Daggett, P.-M., Gherna, R., Jong, S., Simione, F. & Hatt, H. (1980). *American Type Culture Collection Methods. Laboratory Manual on Preservation: Freezing and Freeze Drying*. Rockville, Maryland: American Type Culture Collection.

Allsopp, D. (1985). Fungal culture collections for the biotechnology industry. *Industrial Biotechnology* **5**, 2.

Allsopp, D. & Seal, K. J. (1986). *Introduction to Biodeterioration*, 136 pp. London: Edward Arnold.

Batra, L. R. & Iijima, T. (eds) (1984). *Critical Problems for Culture Collections*, 71 pp. Osaka, Japan: Institute for Fermentation.

Cour, I. G., Maxwell, G. & Hay, R. (1979). Tests for bacterial and fungal contaminants in cell cultures as applied at the ATCC. *TCA Manual* **5**, 1157–60.

Dilworth, S., Hay, R. & Daggett, P.-M. (1979). Procedures in use at the ATCC for detection of protozoan contaminants in cultured cells. *TCA Manual* **5**, 1107–10.

Hawksworth, D. L. (1985). Fungus culture collections as a biotechnological resource. *Biotechnology and Genetic Engineering Reviews* **3**, 417–53.

Hay, R. J. (1983). Availability and standardization of cell lines at the American Type Culture Collection: Current status and prospects for the future. In *Cell Culture Test Methods*, STP 810, ed. S. A. Brown, pp. 114–26. Philadelphia: American Society for Testing and Materials.

Jewell, J. E., Workman, R. & Zelenick, L. D. (1976). Moisture analysis of lyophilized allergenic extracts. In *International Symposium on Freeze-Drying of Biological Products. Developments in Biological Standardization* **36**, 181–9.

Kelley, J. (1985). The testing of plastics for resistance to microorganisms. In *Biodeterioration and Biodegradation of Plastics and Polymers* ed. K. J. Seal, pp. 111–24. Cranfield, UK: Cranfield Press.

Kelley, J. & Allsopp, D. (1987). Mould growth testing of materials, components and equipment to national and international standards. *Society of Applied Bacteriology, Technical Series* **23**, Oxford, UK, Blackwell Scientific Publications.

Lavappa, K. S. (1978). Trypsin-Giemsa banding procedure for chromosome preparations from cultured mammalian cells. *TCA Manual* **4**, 761–4.

Macy, M. (1978). Identification of cell line species by isoenzyme analysis. *TCA Manual* **4**, 833–6.

Macy, M. (1979). Tests for mycoplasmal contamination of cultured cells as applied at the ATCC. *TCA Manual* **5**, 1151–5.

May, M. C., Grim, E., Wheller, R. M. & West, J. (1982). Determination of residual moisture in freeze-dried viral vaccines: Karl Fischer, gravimetric thermogravimetric methodologies. *J. Biological Standardization* **10**, 249–59.

8

Organisation of resource centres

B. E. KIRSOP and E. J. DASILVA

8.1 Introduction

Individual resource and information centres provide valuable services to biotechnology, but their role can be substantially enhanced if their activities are effectively co-ordinated. This has been recognised in the past, and a number of committees, federations and networks have been set up for this purpose at the national, regional and international levels. Although the origins and composition of existing organisations differ and their geographical locations are widespread, their common purpose is to support and develop the activities of resource and information centres for the benefit of microbiology.

8.2 International organisation

8.2.1 *World Federation for Culture Collections*

There are fewer difficulties in setting up national and regional co-ordinating mechanisms than international systems, and yet one of the first developments in this area was the formation of the World Federation for Culture Collections (WFCC). In 1962 at a Conference on Culture Collections held in Canada it was recommended that the International Association of Microbiological Societies (IAMS) set up a Section on Culture Collections. The Section was established in 1963. Five years later, at an International Conference on Culture Collections in Tokyo, the formation of the WFCC was proposed and an *ad hoc* committee, together with the Section on Culture Collections, drew up statutes which were agreed at a congress in 1970. Following the conversion of the IAMS to Union status, the WFCC is now a federation of the International Union of Microbiological Societies (IUMS) and an interdisciplinary Commission of the International Union of Biological Sciences (IUBS).

199

The principal objective of the WFCC is to establish effective liaison between persons and organisations concerned with culture collections and the users of the collections both in the developed and developing regions of the world. To achieve this objective a structure of committees has been set up covering patents, postal and quarantine regulations, education, endangered collections and publicity.

Committee on Patent Procedures. The activities of the Committee on Patent Procedures have important implications for biotechnology. The procedures for patenting processes involving the use of microorganisms, animal or plant cells or of genetically manipulated organisms are described in Chapter 6. The various patent regulations existing in different parts of the world present a confusing picture to those wishing to take out patents, and professional guidance is essential. A number of organisations such as the World Intellectual Property Organisation (WIPO) are concerned with the rationalisation of the different systems, and the WFCC's patents committee has acted in an advisory capacity to them, providing microbiological input. Members of the Committee have attended WIPO meetings to advise on the implementation of the Budapest Treaty for the International Recognition of the Deposit of Microorganisms for the Purpose of Patent Procedures. Additionally, they have monitored the functioning of the Treaty and provided evidence of difficulties that have arisen in its implementation.

Committee for Quarantine and Postal Regulations. The Committee for Quarantine and Postal Regulations is similarly in close communication with the relevant postal regulatory bodies, such as the International Postal Union, National Postal Departments and the International Air Transport Association (IATA), and has put forward recommendations for the safe transport of infectious and non-infectious biological material (see Chapter 3). Members of the committee have been able to encourage international collaboration in this area by attending appropriate meetings and providing specialist advice in order to establish mechanisms for the safe transport of biological material throughout the world.

Education Committee. The WFCC is aware of the lack of guidance given to students before finishing university training on the support and services available from the microbial resource centres of the world. A similar lack of general awareness exists among many working microbiologists in industry, research and education. Accordingly, the Education Committee of the WFCC has an on-going programme of activities to increase the amount of information on back-up available from culture collections. Projects include the publication of books, preparation of training videos, advisory leaflets, and the organisation of training courses, scientific symposia and international conferences. This present series of source books is part of the programme of the Education Committee, designed to increase the usefulness of culture collections to those working in biotechnology.

Committee for Endangered Collections. The Committee for Endangered Collections is concerned to protect the microbial and cellular genetic resources of the world. Many of the major culture collections suffer from time to time from financial restrictions or from a change of direction in the interests of the host institute. Smaller collections are often transitory in nature and face difficulties on the retirement or relocation of the curator whose special interest the collection represents. The WFCC believes the conservation of these collections is of prime importance if the cultures and the substantial investment in terms of effort and expertise are not to be irretrievably lost. To enable emergency measures to be taken when difficulties arise, the Committee for Endangered Collections has obtained financial backing to set up a fund for the provision of specialist, short-term support to allow the relocation of such collections to alternative laboratories willing and competent to take them over. The services of this committee may be used to provide advice to microbiologists who have developed collections of unique microorganisms during the course of their work, but who may not have the wish or expertise to maintain them in the long term or the resources to supply cultures to others.

Publicity Committee. The WFCC's Publicity Committee plays a major role in the dissemination of information about the activities of the Federation to the microbiological community. It produces a newsletter at regular intervals and is closely involved with all administrative developments. In particular, it plays an important part in the four-yearly WFCC International Conference and in the preparation of posters for scientific conferences. The editor of the newsletter will consider the publication of appropriate material and welcomes information about meetings, publications and topics of general interest to members. Biotechnologists may use the newsletter as a forum for the discussion of issues – possibly controversial – that are of interest to fellow scientists. Typical of subjects that can usefully be discussed in the columns of the newsletter are questions relating to stable nomenclature of microorganisms, the retention of published strain designations, security measures for the release of potentially dangerous cultures to those unqualified to handle them or the rescue of important genetic resources.

Data centres. In addition to the functions of these Committees, and others set up from time to time as the need arises, the WFCC has sponsored and is responsible for the World Data Center of Collections of Cultures of Microorganisms. The Center has pioneered the collection of data of this kind and has been responsible for the publication of three Directories listing the collections and the species they hold. The Center was originally housed in the University of Queensland's Department of Microbiology in Australia, but in 1986, on the retirement of its founder Director, was transferred to the Life Sciences Division at RIKEN, Tokyo, Japan. The WFCC is also co-sponsor with CODATA and IUMS of the international Microbial Strain Data Network (MSDN) set up to provide a referral system to the numerous data centres developing throughout the world listing microbial strain data. These two important activities of the WFCC, set up with international funding, are further discussed in Chapter 2.

The WFCC plays a prime role in the organisation of culture collection activities internationally and has among its membership experts in many areas of microbiology. It exists to serve both the culture collections and their users and may be used as a powerful interdisciplinary organ of communication between biotechnologists and specialists in other areas of microbiology.

8.2.2   *The MIRCEN Network*

The *UNESCO Courier* of July 1975 carried a feature 'On the road to development – a UNESCO network for applied microbiology'. Therein several mechanisms – conferences, training courses and fellowships – were identified. Since then, as a means towards strengthening the world network, several regional and international initiatives have been built in through the establishment of microbiological resources centres (MIRCENs) (see Table 8.1). These are designed:

(1) to provide the infrastructure for the building of a world network which would incorporate regional and interregional functional units geared to the management, distribution and utilisation of the microbial gene pool;

(2) to strengthen efforts relating to the conservation of micro-

organisms with emphasis on *Rhizobium* gene pools in developing countries with an agrarian base;
(3) to foster the development of new inexpensive technologies that
are native to the region;
(4) to promote the applications of microbiology in the strengthening of rural economies;

Table 8.1. *Microbial resource centres*

Biotechnology MIRCENs
Ain Shams University, Faculty of Agriculture, Shobra–Khaima, Cairo, Arab
Republic of Egypt

Applied Research Division, Central American Research Institute for Industry
(ICAITI), Ave, La Reforma 4–47 Zone 10, Apdo Postal 1552, Guatemala

CAB International Mycological Institute, Mycology MIRCEN, Ferry Lane, Kew,
Surrey TW9 3AF UK.

Department of Bacteriology, Karolinska Institutet, Fack, S–10401 Stockholm,
Sweden

Fermentation, Food and Waste Recycling MIRCEN, Thailand Institute of Scientific and Technological Research, 196 Phahonyothin Road, Bangken, Bangkok
9, Thailand

Fermentation Technology MIRCEN, ICME, University of Osaka, Suita-shi 656,
Osaka, Japan

Institute for Biotechnological Studies, Research and Development Centre,
University of Kent, Canterbury CT2 7TD, UK.

Marine Biotechnology MIRCEN, Department of Microbiology, University of
Maryland, College Park Campus, Maryland 207742, USA

Planta Piloto de Procesos Industriales Microbiologicos (PROIMI), Avenida
Belgrano y Pasaje Caseros, 4000 S.M. de Tucuman, Argentina

University of Waterloo, Ontario, Canada N2LK 3GI, and University of Guelph,
Guelph, Ontario NIG 2WI, Canada

Rhizobium MIRCENs
Cell Culture and Nitrogen-Fixation Laboratory, Room 116, Building 011-A,
Barc–West, Beltsville, Maryland 20705, USA

Centre National de Recherches Agronomiques, d'Institut Senégalais de
Recherches Agricoles, B.P. 51, Bambey, Senegal

Departments of Soil Sciences and Botany, University of Nairobi, PO Box 30197,
Nairobi, Kenya

IPAGRO, Postal 776, 90000 Porto Alegre, Rio Grande do Sul, Brazil

NifTAL Project, College of Tropical Agriculture and Human Resources, University of Hawaii, PO Box 'O', Paia, Hawaii 96779, USA

World Data Center MIRCEN
World Data Center on Collections of Microorganisms, RIKEN, 2–1 Hirosawa,
Wako, Saitama 351–01, Japan

---

(5) to serve as focal centres for the training of manpower and the imparting of microbiological knowledge.

The first development in the UNESCO global network of Microbiological Resource Centres, consisting of centres in the developed world and regional networks in the developing countries, was the establishment of the World Data Center (WDC) on Microorganisms (see above and Chapter 2) in Queensland, Australia.

The MIRCEN at the Karolinska Institute, Sweden, in addition to developing microbiological techniques for the identification of microorganisms at the WDC, has pioneered the organisation of a series of MIRCENET Computer Conferences on biogas production, anaerobic digestion and the bioconversion of lignocellulose. Computer conferencing is a network system that links geographically scattered nodes together through the use of home or office computers to a remote control computer (see Chapter 2).

Apart from attempting to link up the MIRCENs and organising specialised conferences, MIRCENET has other functions listed in Table 8.2.

On the basis of their research and training programmes, the other MIRCENs can be broadly classified as follows.

The Biotechnology MIRCENS. In the area of biotechnology, there are nine MIRCENs in operation (see Table 8.1). These are in Thailand, Egypt, Guatemala, Japan, Argentina, USA, the UK, Canada and Sweden.

In the region of Southeast Asia, the MIRCEN in Bangkok has co-operating laboratories in the Philippines, Indonesia, Singapore, Malaysia and Hong Kong and other institutions in Thailand. It serves the microbiological community in the collection, preservation, identification and distribution of microbial germplasm, and in the promotion of research and training activities directed towards the needs of the region.

Table 8.2. *Functions of MIRCENET*

To help initiate closed computer conferences under defined keys such as microbiology, biological nitrogen fixation, biogas, networking in culture collections.

To act as an information source for meetings, reviews, identification services, etc.

To provide a platform for discussions on MIRCEN network activities.

To provide print-outs and records of MIRCENET entries.

In the region of the Arab States, the MIRCEN at Ain-Shams University, Cairo, promotes research and training courses on the conservation of microbial cultures and biotechnologies of interest to the region. Through its co-operating MIRCEN laboratory at the University of Khartoum, the MIRCEN has contributed to the establishment of a culture collection in Sudan specialising in fungal taxonomy. The co-operating MIRCEN laboratory at the Institute Agronomique et Veterinaire Hassan II, Rabat, has made commendable progress through projects using different species of yeasts and rhizobia.

In the region of Central America and the Caribbean, the MIRCEN [co-operating laboratories in Chile, Columbia, Costa Rica, Dominican Republic, Ecuador, El Salvador, Honduras, Jamaica, Mexico, Nicaragua, Peru, Venezuela] has, in co-operation with the Organization of American States, the Interamerican Development Bank and several other prestigious agencies, pioneered the applications of microbiology, process engineering and fermentation technology in several member states of Central America and the Caribbean. It has set up joint collaborative research projects, the exchange of technical personnel, regional training programmes and the dissemination of scientific information among network institutions.

The South American Biotechnology MIRCEN located at Tucuman, Argentina is comprised of a regional network with co-operating laboratories in Brazil, Chile, Bolivia and Peru. It has similar goals as the Biotechnology MIRCEN for Central America and the Caribbean.

The MIRCENs in the industrialised societies function as a bridge with those in the developing countries. In such manner, increased co-operation is promoted between the developed and developing countries. Furthermore, a basic structure is set up for eventual twinning at a later date. For example, the Guelph Waterloo MIRCEN, Canada, with its expertise at the University of Waterloo in biomass conversion technology, microbial biomass protein production and bioreactor design, is of immense benefit to the work of the MIRCENs at Cairo, Guatemala and Tucuman.

In a similar manner, the MIRCEN at Bangkok has several collaborative research projects with that at the International Centre of Co-operative Research in Biotechnology, Osaka, Japan. This centre conducts the annual UNESCO International Postgraduate University Course on Microbiology (of 12 months' duration). It also functions as the Japanese point-of-contact for the Southeast Asian regional network of microbiology in the UNESCO Programme for Regional Co-operation in the basic sciences.

In the UK there is a MIRCEN network centred upon the Institute for Biotechnological Studies (IBS). In common with other Microbiological Resource Centres, the aims of the UK MIRCEN are to promote the utilisation of the microbial gene pool, to promote applied microbiology and biotechnology in the developing countries and to provide a centre for training and advice.

The intergovernmental CAB International Mycological Institute (CMI) is the MIRCEN for mycology co-operating with all others in the field world-wide. This organisation, together with the Institute of Horticultural Research (IHR) are the first two organisations to collaborate with the MIRCEN Network, whilst continuing their own activities in mycological and biodeterioration studies, and microbiological pest control, mycorrhizas and mushroom technology respectively.

In keeping with the new trends of the expanding frontiers of biotechnological research, a Marine Biotechnology MIRCEN has been established at the University of Maryland. Work presently underway includes the fundamental elucidation of the evolution of genes and the flow of genes through populations in the marine environment. One collaborative study underway is with the Chinese University of Hong Kong and the Shandong College of Oceanography, Qingdao, China.

The Biological Nitrogen Fixation (BNF) MIRCENS. In the quest for more food for their increasing populations, several developing nations have been expanding their agricultural lands into areas which are marginally capable of sustaining productivity and invariably limited by the availability of nitrogen fertilizer.

In interaction with other international programmes, modest schemes for the development of biofertilizers or *Rhizobium* inoculant material, particularly in legume-crop areas of the developing countries, are already operating through the MIRCENs on a level of regional co-operation in Latin America, East Africa and Southeast Asia and the Pacific.

In the area of biological nitrogen five MIRCENs are already operating (see Table 8.1).

The broad responsibilities of these MIRCENs include collection, identification, maintenance, testing and distribution of rhizobial cultures compatible with crops of the regions. Deployment of local rhizobia inoculant technology and promotion of research are other activities. Advice and guidance are provided in the region to individuals and institutions engaged in rhizobiology research.

The BNF MIRCENs play a valuable role in maintaining and distributing efficient cultures of *Rhizobium*. Nearly 4000 strains are maintained in the MIRCEN collections and about 1750 have been distributed to other organisations (Table 8.3).

The MIRCEN network is founded on the principle of self-help and mutual co-operation. It concentrates on existing facilities and resources and provides an organisational structure which allows each institution to collaborate as best it can through the following:

(1) an exchange of research workers between national and regional institutions;

Table 8.3. *Culture collection services of Biological Nitrogen Fixation (BNF) MIRCENs*

Holdings of Rhizobium *culture collections*

MIRCEN	Number of strains held
Bambey	50
Beltsville	938
Hawaii	2000
Nairobi	208
Porto Alegre	650
Total	3846

Cultures distributed by Rhizobium *MIRCENs*

MIRCEN	Number of cultures	Countries of recipient institutions
Bambey	8	Gambia, Mali, Yemen
Beltsville	508	Zimbabwe, Nigeria, Yugoslavia, India, Spain, Vietnam, Ireland, UK, Malaysia, Italy, Canada, South Africa, Senegal, Egypt, Poland, Argentina, Turkey, W. Germany, Austria, Australia, New Zealand
Hawaii	200	Global
Nairobi	95	Uganda, Malawi, Tanzania, Mauritius, Sudan, Congo, Zaire, Rwanda
Porto Alegre	943	Argentina, Chile, Bolivia, Uruguay, Peru, Ecuador, Columbia, Venezuela, El Salvador, Dominican Rep., Mexico, USA, Trinidad, Brazil

(2) small grants to individual research projects or workers for acquisition of supplies, spare parts for equipment, or small-scale equipment;

(3) participation of senior scientists in specialised symposia in the technically advanced countries in the vicinity of each of the regions;

(4) organisation of short-term intensive training courses and specialised in-depth sub-national or national meetings;

(5) production of a newsletter functioning as an outlet for the exchange of research news, publication of research findings and as an attraction for potential participating laboratories.

The MIRCENs play a catalytic role in breaching the barrier of geographical isolation and advancing the frontiers of contemporary research in biotechnology through the production of newsletter bulletins, culture collection catalogues and research papers. The publication of *MIRCEN News* annually, the development of MIRCENET, and the UNESCO *MIRCEN Journal of Applied Microbiology and Biotechnology* are indications of the gradual emergence of competence and capability of the MIRCENs, and the services they provide on a regional and inter-regional basis.

8.3   Regional organisation

Transnational co-ordinating mechanisms are being set up throughout the world to bring regional cohesion to culture collection activities and benefit to both the resource centres themselves and their users. Some have been established as committees by culture collections; others have originated as data centres with the secondary effect of stimulating closer working collaboration between the contributing culture collections. They may be contacted for information about microbiological resources, services and general advice.

8.3.1   *European Culture Collections' Organisation (ECCO)*

In 1981, at an international conference in Brno, Czechoslovakia, curators of European service culture collections present agreed that a mechanism should be set up to enable meetings to take place on an annual basis for the exchange of ideas and the discussion of common problems. In 1982 the first meeting of ECCO took place at the Deutsche Sammlung von Mikroorganismen, Göttingen, FDR, and since then meetings have been held in France, the UK and Czechoslovakia. Membership has increased steadily as new culture collections are formed or developed to provide a national service. Membership is restricted to collections that provide a service on demand and without restriction, that have as a normal part of their duty the acceptance of cultures, that issue from time to time a list of their holdings, and that are in a country with a microbiological society belonging to the Federation of European Microbiological Societies. It was felt that these collections have interests and problems in common that are not shared by research or teaching collections.

Apart from the exchange of scientific information relating to such topics as taxonomy, identification and preservation procedures, much benefit has been derived from discussions on new developments within culture collections such as acceptance of International Depositary Authority Status (see Chapter 6) or the development of computerised systems for the storage, searching and dissemination of culture information (see Chapter 2). In addition, the opportunity to meet on a regular basis has enabled collaborative programmes to be set up between collections from different countries.

ECCO members have become aware that the services available from the collections are not fully exploited by users. To remedy this they have combined to produce publicity material in the form of brochures, leaflets and scientific posters and plan to set up a permanent information centre in the future. Information about the holdings and services of ECCO collections is available from the Organisation's Officers (see Chapter 2) or the Secretary of FEMS.

8.3.2 *Regional database systems*

A number of co-ordinating mechanisms based on information centres have been set up, primarily to establish data banks for regional access. These are described in greater detail in Chapter 2.

Some, such as the Tropical Data Base in Brazil, the Microbial Information Network in Europe (MINE) and the Nordic Register, have been developed initially to provide a centre for information on the culture collections themselves, their services and their holdings. Others, such as the Microbial Culture Information Service (MiCIS), have been set up in areas with well-established culture collection systems with the purpose of providing on-line strain database for searching. Both these kinds of data centres have the secondary effect of encouraging collaboration between culture collections so that the best possible system develops and minimum duplication of effort takes place.

The proliferation of microbial data centres world-wide reflects the growing need for information on biological materials. It has also led to problems in identifying the most appropriate point of contact for specific information. To overcome this an international system has been established (Microbial Strain Data Network, MSDN), to act as a referral system and communications network to databases able to answer specific enquiries on strain properties (Chapter 2). It seems certain that other systems will be established, and the function of the MSDN thus becomes of increasing importance as the first point of enquiry, directing those seeking information on strain properties to appropriate centres.

8.4   National federations/committees

The following countries have established federations or committees for the co-ordination of culture collection activities.

Australia
Canada
China
Czechoslovakia
Japan
Korea
New Zealand
Turkey
United Kingdom
United States of America

Information about them and their activities may be obtained through culture collections or microbiological societies within the country or through the World Data Center and the Microbial Strain Data Network (see Chapter 2). Most of these organisations produce newsletters from time to time and further information may be obtained through these publications.

Some of the organisations are *for* culture collections, others are *of* culture collections and the difference between the two categories is significant. Those that are *of* culture collections exist primarily to co-ordinate culture collection activities within the countries (produce common catalogues, rationalise holdings, stabilise funding) and are generally termed Committees rather than Federations; those that are *for* culture collections have as their prime function the promotion of communication between the collections and their users in industry, research and education. The activities of the latter category concentrate more on scientific meetings, workshops and training courses, and the membership includes any microbiologists with an interest in culture collection activities, whether they are working in culture collections or not. The executive boards are deliberately formed of people both from culture collections and from research or teaching laboratories and industry, providing a cross fertilisation of interests, whereas with organisations set up *for* culture collections the officers and members are drawn from the collections only. The impact of biotechnological input to the Federations has played a valuable part in the development of microbial resource centres to meet the growing needs of industry in this area.

A number of international organisations exist for the co-ordination of activities within different microbiological disciplines, and information about them can be obtained from the International Council for Scientific Unions (Table 8.4). Information on biotechnology is disseminated through the different associations listed in Table 8.5. All these organisations recognise the need for an effective network of microbial resource centres and are active in support of their development.

8.5 Future developments

Developments in biotechnology have coincided with extensive advances in computer technology, and throughout the world culture collections have taken advantage of the latter to respond to the increasing demands of the former. It is clear from Chapter 2 that data held in the microbial resource centres is increasingly computerised and it is evident that the biotechnology community can better be served by co-ordination of these activities. The World Data Center for Collections of Cultures of Microorganisms and the Microbial Strain Data Network are important examples of international collaboration in this area, leading to on-line databases and information network systems. The imaginative and

Table 8.4. *International scientific organisations*

ICSU	International Council of Scientific Unions 51 Boulevard de Montmorency 75016 Paris France Telephone: 45.25.03.29 Telex: ICSU 630553 F
IUBS	International Union of Biological Sciences
IUMS	International Union of Microbiological Societies
ICRO	(International Cell Research Organisation) Panel on Applied Microbiology and Biotechnology

successful MIRCEN network will continue to be instrumental in encouraging the establishment and development of culture collection activities in the developing world and linking them to those in industrial nations.

Computers will be used increasingly for computer conferencing and electronic mail, leading to greater communication between the collections. This in turn should lead to greater collaborative research and joint service activities and will minimise unnecessary duplication of effort, consistant with national requirements.

In spite of rapid developments in communication systems, the need

Table 8.5. *Biotechnology associations*

AABB	Association for the Advancement of British Biotechnology 1 Queen's Gate London SW1H 9BT UK
ABA	Australian Biotechnical Association 1 Lorraine Street Hampton Victoria 3188 Australia
ABC	Association of Biotechnology Companies 1220 L Street NW Suite 615 Washington, DC 20005 USA
ADEBIO	Association de Biotechnologie 3 rue Massenet 77300 Fontainebleau France
BIDEC	C/o Japan Association of Industrial Fermentation 20–5 Shinbashi 5-chome Minato-ku Tokyo 105 Japan
IBA	Industrial Biotechnology Association 2115 East Jefferson Street Rockville Maryland 20852 USA
IBAC	Industrial Biotechnology Association of Canada Lava University Cité Universitaire Quebec G1K 7P4 Canada

for the presence of culture collections in all regions of the world will remain because of specialised local needs, regional regulatory requirements, such as those for postal and quarantine purposes, or currency or language reasons. Duplication of important holdings and services is necessary, but can be reduced to an acceptable level by collaborative efforts on the part of individual scientists in the resource centres, the setting up of organisations to co-ordinate their activities and the use of computers and electronic networking to facilitate communication. A basic core of collaborative mechanisms already exists and can be extended to cover regions of the world or specialist areas of activity not yet co-ordinated internationally.

REFERENCES

In re Abitibi (1982). *Canadian Patent Reporter* **62**, 81.

Adoutte-Panvier, A., Davies, J. E., Gritz, L. R. & Littlewood, B. S. (1980). Studies of ribosomal proteins of yeast species and their hybrids: Gel electrophoresis and immunochemical crossreactions. *Mol. Gen. Genet.* **179**, 273–82.

Alexander, M. T. & Simione, F. (1980). Factors affecting the recovery of freeze-dried *Saccharomyces cerevisiae*. American Society of Microbiology Abstracts of Annual Meeting 1980.

Alexander, M., Daggett, P. -M., Gherna, R., Jong, S., Simione, F. & Hatt, H. (1980). American Type Culture Collection Methods. *Laboratory Manual on Preservation: Freezing and Freeze-Drying*. Rockville, Maryland: American Type Culture Collection.

Ando, T., Shibata, T. & Watabe, H. (1984). Endo-deoxyribonuclease and process for the production thereof. US Patent 4,430,432.

Annear, D. I. (1958). Preservation of microorganisms by drying from the liquid state. In *Proceedings of the First International Conference for Culture Collections*, ed. H. Iijuka & T. Hasegawa, pp. 273–6. Tokyo: University of Tokyo Press.

Aulakh, H. S., Straus, S. E. & Kwon-Chung, K. J. (1981). Genetic relatedness of *Filobasidiella neoformans (Cryptococcus neoformans)* and *Filobasidiella bacillispora (Cryptococcus bacillisporus)* as determined by deoxyribonucleic acid base composition and sequence homology studies. *Int. J. Syst. Bacteriol.* **31**, 97–103.

Bahareen, S., Melcher, U. & Vishniac, H. S. (1983). Complementary DNA-25S ribosomal RNA hybridization: An improved method for phylogenetic studies. *Can. J. Microbiol.* **29**, 546–51.

Baker, J. G., Salkin, I. F., Pincus, D. H. & D'Amato, R. F. (1981). Use of rapid auxanographic procedures for recognition of an atypical *Candida*. *J. Clin. Microbiol.* **13**, 652–4.

Ballou, C. E. (1974). Some aspects of the structure, immunochemistry, and genetic control of yeast mannans. *Adv. Enzymol. Relat. Areas Mol. Biol.* **40**, 239–70.

214

Banno, I., Mikata, K. & Yamauchi, S. (1981) Preservation of yeast cultures on anhydrous silica gel. *Institute for Fermentation, Osaka, Research Communications* **10**, 39–44.

Baptist, J. N. & Kurtzman, C. P. (1976). Comparative enzyme patterns in *Cryptococcus laurentii* and its taxonomic varieties. *Mycologia* **68**, 1195–203.

Baptist, J. N., Shaw, C. R. & Mandel, M. (1971). Comparative zone electrophoresis of enzymes of *Pseudomonas solanacearum* and *Pseudomonas cepacia. J. Bacteriol.* **108**, 799–803.

Barnett, J. A., Payne, R. W. & Yarrow, D. (1983). *Yeasts, characteristics and identification.* Cambridge University Press.

Bassel, J., Contopoulou, R., Mortimer, R. & Fogel, S. (1977). In *UK Federation for Culture Collections Newsletter*, No. 4, p. 7.

Baumann, L., Bang, S. S. & Baumann, P. (1980). Study of relationship among species of *Vibrio, Photobacterium*, and terrestrial enterobacteria by an immunological comparison of glutamine synthetase and superoxide dismutase. *Curr. Microbiol.* **4**, 133–8.

Baumann, L. & Baumann, P. (1978). Studies of relationship among terrestrial *Pseudomonas, Alcaligenes*, and enterobacteria by an immunological comparison of glutaime synthetase. *Arch. Microbiol.* **119**, 25–30.

Beech, F. W. & Davenport, R. R. (1971). Isolation, purification and maintenance of yeasts. In *Methods in Microbiology*, vol. 4, ed. C. Booth, pp. 153–82. London: Academic Press.

Beech, F. W. *et al.* (1980). Media and methods for growing yeasts: Proceedings of a discussion meeting. In *Biology and Activities of Yeasts*, ed. F. A. Skinner, S. M. Passmore & R. R. Davenport, pp. 259–301. London: Academic Press.

Beier, F. K., Crespi, R. S. & Straus, J. (1985). *Biotechnology and Patent Protection: An International Review.* Paris: Organization for Economic Co-operation and Development.

Bicknell, J. N. & Douglas, H. C. (1970). Nucleic acid homologies among species of *Saccharomyces. J. Bacteriol.* **101**, 505–12.

Blanz, P. A. & Gottschalk, M. (1986). Systematic position of *Septobasidium, Graphiola* and other basidiomycetes as deduced on the basis of their 5S ribosomal RNA nucleotide sequences. *System. Appl. Microbiol.* **8**, 121–7.

Bothast, R. J., Kurtzman, C. P., Saltarelli, M. D. & Slininger, P. J. (1986). Ethanol production by 107 strains of yeasts on 5, 10, and 20% lactose. *Biotech. Letters* **8**, 593–6.

Bowman, P. I. & Ahearn, D. G. (1976). Evaluation of commercial systems for the identification of clinical yeast isolates. *J. Clin. Microbiol.* **4**, 49–53.

Brodelius, P., Nilsson, K. & Mosbach, K. (1981). Production of α-keto acids with alginate-entrapped whole cells of the yeast *Trigonopsis variabilis*. In *Advances in Biotechnology*, vol. III, *Fermentation Products*, ed. C. Vezina & K. Singh, pp. 373–6. Toronto: Pergamon Press.

Budapest Treaty (1981). *Budapest Treaty on the International Recognition of the Deposit of Microorganisms for the Purposes of Patent Procedure 1977* and *Regulations 1981.* Geneva: World Intellectual Property Organization.

Butterfield, W. & Jong, S. C. (1975). Retention of albino and brown phenotypes of *Histoplasma capsulatum* by liquid nitrogen refrigeration. *Sabouraudia* **14**, 291–4.

Calcott, P., Wood, D. & Anderson, L. (1983). Freezing and thawing induced curing of drug-resistance plasmids from bacteria. *Cryo-Letters* **4**, 99–106.

Campbell, I. (1973). Numerical analysis of *Hansenula, Pichia*, and related yeast genera. *J. Gen. Microbiol.* **77**, 427–41.

Convention (1980). *Convention of the Grant of European Patents.* (European Patent Convention) 1973 with 1980 amendments. Munich: European Patent Organization.

Cooney, C. L. & Schaefer, E. J. (1982). Process for producing maltase. US Patent 4,332,899.

Crespi, R. S. (1982). *Patenting in the Biological Sciences.* Chichester: John Wiley & Sons.

Crespi, R. S. (1985). Patent protection in biotechnology: questions, answers and observations. In *Biotechnology and Patent Protection*, ed. F. K. Beier, R. S. Crespi & J. Straus, pp. 36–85. Paris: Organization for Economic Co-operation and Development.

Demain, A. L. (1981). Industrial microbiology. *Science* **214**, 987–95.

De Zeeuw, J. R. & Tynan, E. J., III (1973). Fermentation process for the production of D-mannitol. US Patent 3,736,229.

In re Diamond & Chakrabarty (1980). *US Patents Quarterly 206*, 193.

Duffy, J. I. (1980). *Chemicals by Enzymatic and Microbial Processes. Recent Advances.* Park Ridge, New Jersey: Noyes Data Corp.

du Preez, J. C. & J. P. van der Walt. (1983). Fermentation of D-xylose to ethanol by *Candida shehatae. Biotechnol. Lett.* **5**, 357–62.

EBC Analytica Microbiologica (1977). *Journal of the Institute of Brewing* **83**, 109–18.

Fenton, D. M. (1982). Lactase preparation. US Patent 4,329,429.

Ferenczy, L. (1981). Microbial protoplast fusion. In *Genetics as a Tool in Microbiology*, Symposium of the Society of General Microbiology, vol. 31, ed. S. W. Glover & D. A. Hopwood, pp. 1–34. London & New York: Cambridge University Press.

Fox, G. E., Stackebrandt, E., Hespell, R. B., Gibson, J., Maniloff, J., Dyer, T. A., Wolfe, R. S., Balch, W. E., Tanner, R. S., Magrum, L. J., Zablen, L. B., Balkemore, R., Gupta, R., Bonen, L., Lewis, B. J., Stahl, D. A., Leuhrsen, K. R., Chen, K. N. & Woese, C. R. (1980). The phylogeny of prokaryotes. *Science* **209**, 457–63.

Fry, R. M. (1966). Freezing and drying of bacteria. In *Cryobiology*, ed. H. T. Meryman pp. 665–96. London & New York: Academic Press.

Fujiwara, A. & Masuda, S. (1981). Process for producing D-arabitol. US Patent 4,271,268.

Fukumura, T. (1976). Hydrolysis of L-α-amino-ε-caprolactam by yeasts. *Agr. Biol. Chem.* **40**, 1695–8.

Fuson, G. B., Price, C. W. & Phaff, H. J. (1979). Deoxyribonucleic acid sequence relatedness among some members of the yeast genus *Hansenula. Int. J. Syst. Bacteriol.* **29**, 64–9.

Fuson, G. B., Price, C. W. & Phaff, H. J. (1980). Deoxyribonucleic acid base sequence relatedness among strains of *Pichia ohmeri* that produce dimorphic ascospores. *Int. J. Syst. Bacteriol.* **30**, 217–19.

Hagler, A. N. & Ahearn, D. G. (1981). A rapid DBB test to detect basidiomycetous affinity of yeasts. *Int. J. Syst. Bacteriol.* **31**, 204–8.

Hesseltine, C. W. (1983). Microbiology of Oriental fermented foods. *Ann. Rev. Microbiol.* **37**, 575–601.

Hieda, K. & Ito, T. (1973). Induction of genetic change by drying in yeast. In *Freeze-drying of Biological Materials – Proceedings of C-1 Symposium (Sapporo.)*, pp. 71–8. Paris: International Institute of Refrigeration.

Holzschu, D. L. (1981). 'Molecular taxonomy and evolutionary relationships among cactophilic yeasts.' Ph.D. Thesis, University of California, Davis.

Hubalek, Z. & Kochová-Kratochvilová, A. (1978). Liquid nitrogen storage of yeast cultures. 1. Survival and literature review of the preservation of fungi at ultralow temperatures. *Antonie van Leeuwenhoek* **44**, 229–41.

International Convention (1978). *International Convention for the Protection of New Varieties of Plants* 1961, revised 1972. Geneva: World Intellectual Property Organization.

Jarl, K. (1969). Symba yeast process. *Food Technol.* **23**, 1009–12.

Johnson, J. L. (1981). Genetic characterization. In *Manual of Methods for General Bacteriology*, ed. P. Gerhardt, pp. 450–72. Washington, DC: American Soc. Microbiol.

Johnson, J. L. & Harich, B. (1983). Comparisons of procedures for determining ribosomal ribonucleic acid similarities. *Curr. Microbiol.* **9**, 111–20.

Kaneko, Y., Mikata, K. & Banno, I. (1985). Maintenance of recombinant plasmids in *Saccharomyces cerevisiae* after L-drying. *Institute for Fermentation, Osaka, Research Communications* **12**, 78–82.

Kennell, D. E. (1971). Principles and practices of nucleic acid hybridization. *Prog. Nucleic Acid Res. Mol. Biol.* **11**, 259–301.

Kikuchi, T., Ogawa, M. & Ando, M. (1983). Process for producing acyl-coenzyme A oxidase. US Patent 4,371,620.

Kirsop, B. E. (1974). The stability of biochemical, morphological and brewing properties of yeast cultures maintained by subculturing and freeze-drying. *Journal of the Institute of Brewing* **80**, 565–70.

Kirsop, B. E. (1978). In *Abstracts of the XII International Congress of Microbiology, 1978*, München, p. 39.

Kirsop, B. E. (1984). Maintenance of yeasts. In *Maintenance of Microorganisms*, ed. B. E. Kirsop & J. J. S. Snell, Chapter 12, pp. 109–30. London: Academic Press.

Kirsop, B. E. & Henry, J. E. (1984). Development of a miniaturised cryopreservation method for the maintenance of a wide range of yeasts. *Cryo-Letters* **5**, 191–200.

Kirsop, B. E., Painting, K. A. Henry, J. E., Fernandes, M., Prescott, E. H., Braid, I. & Foster, P. (1986). On-line computer assisted identification of yeasts. *XIV International Congress of Microbiology Abstracts*, p. 75.

Kockova-Kratochvilova, A. & Blagodatskaja, V. (1974). The evaluation of freeze-dried yeasts. *Biologia (Bratislava)* **29**, 893–901.

Kraepelin, G. & Schulze, U. (1982). *Sterigmatosporidium* gen. n., a new heterothallic basidiomycetous yeast, the perfect state of a new species of *Sterigmatomyces* Fell. *Antonie van Leeuwenhoek* **48**, 471–83.

Kreger-van Rij, N. J. W. (1977). Electron microscopy of sporulation in *Schwanniomyces alluvius*. *Antonie van Leeuwenhoek* **43**, 55–64.

Kreger-van Rij, N. J. W. (1984). *The Yeasts, A Taxonomic Study*. Amsterdam: Elsevier.

Kreger-van Rij, N. J. W. & Veenhuis, M. (1971). A comparative study of the cell wall structure of basidiomycetous and related yeasts. *J. Gen. Microbiol.* **68**, 87–95.

Kreger-van Rij, N. J. W. & Veenhuis, M. (1973). Electron microscopy of septa in ascomycetous yeasts. *Antonie van Leeuwenhoek* **39**, 481–90.

Kurtzman, C. P. (1984a). Resolution of varietal relationships within the species *Hansenula anomala, Hansenula bimundalis,* and *Pichia nakazawae* through comparisons of DNA relatedness. *Mycotaxon* **19**, 271–9.

Kurtzman, C. P. (1984b). Synonomy of the yeast genera *Hansenula* and *Pichia* demonstrated through comparisons of deoxyribonucleic acid relatedness. *Antonie van Leeuwenhoek* **50**, 209–17.

Kurtzman, C. P., Johnson, C. J. & Smiley, M. J. (1979). Determination of conspecificity of *Candida utilis* and *Hansenula jadinii* through DNA reassociation. *Mycologia* **71**, 844–7.

Kurtzman, C. P. & Kreger-van Rij, N. J. W. (1976). Ultrastructure of ascospores from *Debaryomyces melissophilus*, a new taxonomic combination. *Mycologia* **68**, 422–5.

Kurtzman, C. P., Phaff, H. J. & Meyer, S. A. (1983). Nucleic acid relatedness among yeasts. In *Yeast Genetics, Fundamental and Applied Aspects*, ed. J. F. T. Spencer, D. M. Spencer & A. R. W. Smith, pp. 139–66. New York: Springer-Verlag.

Kurtzman, C. P. & Smiley, M. J. (1974). A taxonomic re-evaluation of the round-spored species of *Pichia*. In *Proceedings of the Fourth International Symposium on Yeasts*, Vienna, Austria, Part I, ed. H. Klaushofer & U. B. Sleytr, pp. 231–2. Vienna: Hochschulerschaft an der Hochschule für Bodenkultur.

Kurtzman, C. P. & Smiley, M. J. (1979). Taxonomy of *Pichia carsonii* and its synomyms *Pichia vini* and *P. vini* var. *melibiosi*: comparison by DNA reassociation. *Mycologia* **71**, 658–62.

Kurtzman, C. P., Smiley, M. J. & Baker, F. L. (1972). Scanning electron microscopy of ascospores of *Schwanniomyces*. *J. Bacteriol.* **112**, 1380–2.

Kurtzman, C. P., Smiley, M. J. & Baker, F. L. (1975). Scanning electron microscopy of ascospores of *Debaryomyces* and *Saccharomyces*. *Mycopathol. Mycol. Appl.* **55**, 29–34.

Kurtzman, C. P., Smiley, M. J. & Johnson, C. J. (1980a). Emendation of the genus *Issatchenkia* Kudriavzev and comparison of species by deoxyribonucleic acid reassociation, mating reaction, and ascospore ultrastructure. *Int. J. Syst. Bacteriol.* **30**, 503–13.

Kurtzman, C. P., Smiley, M. J., Johnson, C. J. & Hoffman, M. J. (1980b). Deoxyribonucleic acid relatedness among species of *Sterigmatomyces*. *Abstr. Int. Symp. Yeasts*, 5th Y-5.2.5(L), p. 246.

Kurtzman, C. P., Smiley, M. J., Johnson, C. J., Wickerham, L. J. & Fuson, G. B. (1980c). Two new and closely related heterothallic species, *Pichia amylophila* and *Pichia mississippiensis*: Characterization by hybridization and deoxyribonucleic acid reassociation. *Int. J. Syst. Bacteriol.* **30**, 208–16.

Kurtzman, C. P., Vesonder, R. F. & Smiley, M. J. (1973). Formation of extracellular C_{14}–C_{18} 2-D-hydroxy fatty acids by species of *Saccharomycopsis*. *Appl. Microbiol.* **26**, 650–2.

Kurtzman, C. P., Vesonder, R. F. & Smiley, M. J. (1974). Formation of extracellular 3-D-hydroxypalmitic acid by *Saccharomycopsis malanga* comb. nov. *Mycologia* **66**, 580–7.

Lachance, M. A. & Phaff, H. J. (1979). Comparative study of molecular size and structure of exo-β-glucanases from *Kluyveromyces* and other yeast genera: evolutionary and taxonomic implications. *Int. J. Syst. Bacteriol.* **29**, 70–8.

Lapage, S. P., Bascomb, S., Willcox, W. R. & Curtis, M. A. (1973). Identification of bacteria by computer: General aspects and perspectives. *J. Gen. Microbiol.* **77**, 273–90.

Leathers, T. D., Kurtzman, C. P. & Detroy, R. W. (1984). Overproduction and regulation of xylanase in *Aureobasidium pullulans* and *Cryptococcus albidus*. *Biotech. Bioeng. Symp.* **14**, 225–40.

Lindegren, C. C. & Lindegren, G. (1949). Unusual gene-controlled combinations of carbohydrate fermentations in yeast hybrids. *Proc. Natl. Acad. Sci. USA* **35**, 23–7.

In re Lundak (1985). *US Patents Quarterly* **227**, 90.

McArthur, C. R. & Clark-Walker, G. D. (1983). Mitochondrial DNA size diversity in the *Dekkera/Brettanomyces* yeasts. *Curr. Genet.* **7**, 29–35.

Marmur, J. & Doty, P. (1962). Determination of the base composition of DNA from its thermal denaturation temperature. *J. Mol. Biol.* **5**, 109–18.

Martini, A. & Phaff, H. J. (1973). The optical determination of DNA–DNA homologies in yeasts. *Ann. Micro.* **23**, 59–68.

Mazur, P. (1970). Cryobiology: The freezing of biological systems. *Science* **168**, 939–49.

Medonça-Hagler, L. C. & Phaff, H. J. (1975). Deoxyribonucleic acid base composition and DNA/DNA hybrid formation in psychrophobic and related yeasts. *Int. J. Syst. Bacteriol.* **25**: 222–9.

Meyer, S. A. & Phaff, H. J. (1972). DNA base composition and DNA–DNA homology studies as tools in yeast systematics. In *Yeasts, Models in Science and Technics*, ed. A. Kochová-Kratochvilová & E. Minarik, pp. 375–86. Bratislava, Czechoslovakia: Publ. House Slovak Acad. Sci.

Meyer, S. A., Smith, M. T. & Simione, Jr. F. P., (1978). Systematics of *Hanseniaspora* Zikes and *Kloeckera* Janke. *Antonie van Leeuwenhoek* **44**, 79–96.

Mikata, K. & Banno, I. (1986). Preservation of yeast cultures by freezing at −80°C. *Japanese Journal of Freezing and Drying* **32**, 58–63.

Mikata, K., Yamauchi, S. & Banno, I. (1983). Preservation of yeast cultures by L-drying. *Institute of Fermentation, Osaka, Research Communication* **11**, 25–46.

Miles, A. A. & Misra, S. S. (1938). The estimation of the bactericidal power of the blood. *Journal of Hygiene, Cambridge* **38**, 732–49.

Morris, G. J. (1981). *Cryobiology*. Cambridge Institute of Terrestrial Ecology.

Murtagh, J. E. (1986). Fuel ethanol production – the US experience. *Proc. Biochem.* **21**, 61–5.

Nakase, T. & Komagata, K. (1968). Taxonomic significance of base composition of yeast DNA. *J. Gen. Appl. Microbiol.* **14**, 345–57.

Peppler, H. J. (1970). Food yeasts. In *The Yeasts*, vol. 3, *Yeast Technology*, ed. A. H. Rose & J. S. Harrison, pp. 421–62. New York: Academic Press.

Phaff, H. J. (1971). Structure and biosynthesis of the yeast cell envelope. In *The Yeasts*, vol. 2, *Physiology and Biochemistry of Yeasts*, ed. A. H. Rose & J. S. Harrison, pp. 135–210. London: Academic Press.

Phaff, H. J., Miller, M. W. & Miranda, M. (1979). *Hansenula alni*, a new heterothallic species of yeast from exudates of alder trees. *Int. J. Syst. Bacteriol* **29**, 60–3.

Price, C. W., Fuson, G. B. & Phaff, H. J. (1978). Genome comparison in yeast systematics: Delimitation of species within the genera *Schwanniomyces, Saccharomyces, Debaryomyces,* and *Pichia. Microbiol. Rev.* **42**, 161–93.

Pyke, M. (1958). The technology of yeast. In *The Chemistry and Biology of Yeasts*, ed. A. H. Cook, pp. 535–86. New York: Academic Press.

Robbins, E. A. & Seeley, R. D. (1978). Process for the manufacture of yeast glycan. US Patent 4,122,196.

Rogosa, M., Krichevsky, M. I. & Colwell, R. R. (1986). *Coding Microbiological Data for Computers*. New York: Springer-Verlag.

Ruffles, G. (1986). Patents and the biologist. *Biologist* **33**, 5–10.

Schildkraut, C. L., Marmur, J. & Doty, P. (1962). Determination of the base composition of deoxyribonucleic acid from its buoyant density in CsCl. *J. Mol. Biol.* **4**, 430–3.

Schneider, H., Wang, P. Y., Chan, Y. K. & Maleszka, R. (1981). Conversion of D-xylose into ethanol by the yeast *Pachysolen tannophilus. Biotechnol. Lett.* **3**, 89–92.

Selander, R. K. (1976). Genetic variation in natural populations. In *Molecular Evolution*, ed. F. J. Ayala, pp. 21–46. Sunderland, Massachusetts: Sinauer.

Slininger, P. J., Bothast, R. J., Vancauwenberge, J. R. & Kurtzman, C. P. (1982). Conversion of D-xylose to ethanol by the yeast *Pachysolen tannophilus. Biotechnol. Bioeng.* **24**, 371–84.

Slodki, M. E. (1980). Structural aspects of exocellular yeast polysaccharides. In *ACS Symposium Series*, No. 126, *Fungal Polysaccharides*, ed. P. A. Sandford & K. Matsuda, pp. 183–96. Washington, DC: American Chemical Society.

Smith, M.T. (1986). *Zygoascus hellenicus* gen. nov., sp. nov., the teleomorph of *Candida hellenica* (= *C. inositophila* = *C. steatolytica). Antonie van Leeuwenhoek* **52**, 25–37.

Smith, M.T., Batenburg-van der Vegte, W. H. & Scheffers, W. A. (1981). *Eeniella*, a new genus of the Torulopsidales. *Int. J. Syst. Bacteriol.* **31**, 196–203.

Souzu, H. (1973). The phospholipid degradation and cellular death caused by freeze-thawing or freeze-drying of yeast. *Cryobiology* **10**, 427–31.

Spencer, J. F. T. & Gorin, P. A. J. (1969). Systematics of the genera *Hansenula* and *Pichia*: Proton magnetic resonance spectra of their mannans as an aid in classification. *Can. J. Microbiol.* **15**: 375–82.

Spencer, J. F. T., Gorin, P. A. J. & Rank, G. H. (1971). The genetic control of the two types of mannan produced by *Saccharomyces cerevisiae*. *Can. J. Microbiol.* **17**, 1451–4.

Stodola, F. H. & Wickerham, L. J. (1960). Formation of extracellular sphingolipids by microorganisms. II. Structural studies on tetraacetylphytosphingosine from the yeast *Hansenula ciferrii*. *J. Biol. Chem.* **235**, 2584–5.

Straus, J. (1985). *Industrial Property Protection of Biotechnological Inventions: Analysis of Certain Basic Issues.* Document BIG/281. Geneva: World Intellectual Property Organization.

Terada, O. (1972). Process for producing lipase. US Patent 3,692,632.

Toivola, A., Yarrow, D., van den Bosch, E., van Dijken, J. P. & Scheffers, W. A. (1984). Alcoholic fermentation of D-xylose by yeasts. *Appl. Environ. Microbiol.* **47**, 1221–9.

Tsuchiya, T., Fukazawa, Y., Taguchi, M., Nakase, T. & Shinoda, T. (1974). Serological aspects of yeast classification. *Mycopathol. Mycol. Appl.* **53**, 77–91.

US Department of Health and Human Services, Public Health Service (1983). *Biosafety in Microbiological and Biomedical Laboratories.* National Institutes of Health, Bethesda, Maryland, USA.

US Department of Health, Education and Welfare (1972). *Classification of Etiologic Agents on the Basis of Hazard.* Centers for Disease Control, Atlanta, Georgia 30333, USA.

van der Walt, J. P. & Hopsu-Havu, V. K. (1976). A colour reaction for the differentiation of ascomycetous and hemibasidiomycetous yeasts. *Antonie van Leeuwenhoek* **42**, 157–63.

van der Walt, J. P. & von Arx, J. A. (1980). The genus *Yarrowia* gen. nov. *Antonie van Leeuwenhoek* **46**, 517–21.

van der Walt, J. P. & Yarrow, D. (1984a). Methods for isolation, maintenance, classification and identification of yeasts. In *The Yeasts, a Taxonomic Study*, ed. N. J. W. Kreger-van Rij, pp. 45–104. Amsterdam: Elsevier Science Publishers B. V.

van der Walt, J. P. & Yarrow, D. (1984b). The genus *Arxiozyma* gen. nov. (Saccharomycetaceae). *S. Afr. J. Bot.* **3**, 340–2.

Vaughan Martini, A. & Kurtzman, C. P. (1985). Deoxyribonucleic acid relatedness among species of the genus *Saccharomyces* sensu stricto. *Int. J. Syst. Bacteriol.* **35**, 508–11.

Vesonder, R. F., Stodola, F. H., Rohwedder, W. K. & Scott, D. B. (1970). 2-D-Hydroxyhexadecanoic acid: A metabolic product of the yeast *Hansenula sydowiorum*. *Can. J. Chem.* **48**, 1985–6.

von Rehberg, R. (1978). Kulturhaltung und Konservierung von Hefestämmen. *Branntweinwirtschaft* **118** (No. 1, January), 1–6.

Walker, W. F. (1985). 5S Ribosomal RNA sequences from ascomycetes and evolutionary implications. *Syst. Appl. Microbiol.* **6**, 48–53.

Walker, W. F. & Doolittle, W. F. (1982). Redividing the basidiomycetes on the basis of 5S rRNA sequences. *Nature* (London) **299**, 723–4.

Walker, W. F. & Doolittle, W. F. (1983). 5S rRNA sequences from eight basidiomycetes and fungi imperfecti. *Nucleic Acids Res.* **11**, 7625–30.

Wellman, A. M. & Stewart, G. G. (1973). Storage of brewing yeasts by liquid nitrogen refrigeration. *Applied Microbiology* **26**(4), 577–83.

White, W. B. & Wharton, K. L (1984). Development of a mechanically refrigerated cryogenic preservation system using a mixed refrigerant technique to maintain specimen temperatures of $-135\,^{\circ}$C. In *American Laboratory* (1984), 1–7.

Wickerham, L. J. (1951). Taxonomy of yeasts. *US Dept. Agric. Tech. Bull.* **1029**, 1–56.

Wickerham, L. J. & Burton, K. A. (1954). A clarification of the relationship of *Candida guilliermondii* to other yeasts by a study of their mating types. *J. Bacteriol.* **68**, 594–7.

Wickerham, L. J. & Burton, K. A. (1962). Phylogeny and biochemistry of the genus *Hansenula*. *Bacteriol. Rev.* **26**, 382–97.

Wickerham, L. J., Flickinger, M. H. & Johnston, R. M. (1946). The production of riboflavin by *Ashbya gossypii*. *Arch. Biochem.* **9**, 95–8.

Wickerham, L. J., Kurtzman, C. P. & Herman, A. I. (1969). Sexuality in *Candida lipolytica*. In *Recent Trends in Yeast Research.* vol. I, ed. D. G. Ahearn, pp. 81–92. Atlanta: Georgia State University.

Wickerham, L. J., Lockwood, L. B., Pettijohn, O. G. & Ward, G. E. (1944). Starch hydrolysis and fermentation by the yeast *Endomycopsis fibuliger*. *J. Bacteriol.* **48**, 413–27.

Wickerham, L. J. & Stodola, F. H. (1960). Formation of extracellular sphingolipids by microorganisms. I. Tetraacetylphytosphingosine from *Hansenula ciferrii*. *J. Bacteriol.* **80**, 484–91.

Winge, O. & Roberts, C. (1949). Inheritance of enzymatic characters in yeast and the phenomenon of long-term adaptation. *Comp. Rend. Trav. Lab. Carlsberg* **24**, 263–315.

WIPO (1980). *Records of the Budapest Diplomatic Conference for the Conclusion of a Treaty on the International Recognition of the Deposit of Microorganisms for the Purposes of Patent Procedure.* WIPO Publication no. 332 (E), p. 119. Geneva: World Intellectual Property Organization.

WIPO (1986). *Report adopted by the 2nd session of the Paris Union Committee of Experts on Biotechnological Inventions and Industrial Property.* WIPO Document BioT/CE/II/3. Geneva: World Intellectual Property Organization.

Yamada, K. (1977). *Japan's Most Advanced Industrial Fermentation Technology and Industry.* Tokyo: International Technological Information Institute.

Yamada, Y. & Nakase, T. (1985). *Waltomyces*, a new ascosporogenous yeast genus for the Q_{10}-equipped, slime-producing organisms whose asexual reproduction is by multilateral budding and whose ascospores have smooth surfaces. *J. Gen. Appl. Microbiol.* **31**, 491–2.

Yamazaki, M. & Komagata, K. (1981). Taxonomic significance of electrophoretic comparison of enzymes in the genera *Rhodotorula* and *Rhodosporidium*. *Int. J. Syst. Bacteriol.* **31**, 361–81.

Yarrow, D. & Meyer, S. A. (1978). Proposal for amendment of the diagnosis of the genus *Candida* Berkhout nom. cons. *Int. J. Syst. Bacteriol.* **28**, 611–15.

INDEX

accession numbers, 64–5
accessions by culture collections, 2–3, 62–5
accreditation of laboratories, 195
Aciculoconidium, characteristics, 113
acyl-coenzyme A oxidase, commercial production, 138
Advisory Committee on Dangerous Pathogens (ACDP), 68
Advisory Committee on Genetic Manipulation, 70, 73
advisory services, on maintenance of organisms, 189
on strain selection, 189
Agricultural Research Service Culture Collection (NRRL) (USA), 4, 13, 15, 16, 32–3, 171, 179–81
air transport, 71
All-Union Collection of Microorganisms (VKM–BKM), 12, 15, 16, 29–30
Ambrosiozyma, characteristics of, 101
dolipores in, 135
American Type Culture Collection (ATCC), 12, 15, 16, 31–2
patent deposits, 171, 178–9
regulations on transportation, 72, 73
supply of pathogens, 70
L-aminolactam, conversion to L-lysine, 138
amylase, commercial production, 137–8
anamorphic (imperfect) yeasts, 113–14, 121, 126, 128
animal varieties, legal protection, 146
animals, processes for production of, patent protection for, 146–7
D-arabitol, 139
Argentina, MIRCEN in, 204, 205, 206
Arthroascus, characteristics, 101

Arxiozyma, characteristics, 101
Ascoidea, 140
ascomycetes, 119
DBB test, 121
genera and characteristics, 101–9
ascospores, 118, 119, 134
Ashbya, 140
Ashbya gossypii, 140
Asia, resource centres, 8–21
Aspergillus niger, 138–9
assimilation tests, 120–1
Association de Biotechnologie, 212
Association for the Advancement of British Biotechnology, 212
Association of Biotechnology Companies, 212
Atlantic Regional Laboratory (ARL) of the National Research Council (Canada), 46
Aureobasidium pullulans, 140
Australia, 45, 210
patent system, 142, 153
resource centres in, 21
Australian Biotechnical Association, 212
Austria, patent deposits, 153
auxanographic growth tests, 121

baking, 1, 136
basidiomycetes, 119
DBB test, 121
general and characteristics, 110–12
basidiospores, 118–19
Bayerische Landesanstalt für Weinbau und Gartenbau (LWG), 11, 14, 16, 26
Belgium, 47, 153
BIDEC (c/o Japan Association of Industrial Fermentation), 212

225